Evolution Driven by Organismal Behavior

Rui Diogo

Evolution Driven by Organismal Behavior

A Unifying View of Life, Function, Form, Mismatches, and Trends

 Springer

Rui Diogo
Department of Anatomy
Howard University College of Medicine
Washington, DC
USA

ISBN 978-3-319-83773-4   ISBN 978-3-319-47581-3   (eBook)
DOI 10.1007/978-3-319-47581-3

Printed on acid-free paper

This Springer imprint is published by Springer Nature
The registered company is Springer International Publishing AG
The registered company address is: Gewerbestrasse 11, 6330 Cham, Switzerland

Never forget that only dead fish swim with the stream.

Thomas Malcolm Muggeridge (1903–1990)

By displaying, every day, a strikingly diverse range of complex—mostly surprising and often flagrantly non-optimal—behavioral choices, a marvelously fascinating friend has led me to shift my previous, mainly internalist, view of life to recognize the crucial importance of behavioral shifts in evolution. I therefore dedicate the book to Tots, the exceptional member of the species *Canis lupus* shown on the cover.

Preface

There is an important point I want to make clear that is key within my philosophy. I am an avid—if not slightly obsessed—reader. I like to read; I wish I could know everything about everything, but unfortunately there is too much to read and so little time. As in any modern book, most of the ideas I focus on in this monograph were therefore not conceived by me for the first time. Accordingly, I profoundly believe that we should not only clearly mention, but actually pay tribute to, those authors who have influenced our ideas either because we agree with them or because by disagreeing with them we re-think and eventually change our original ideas. My viewpoint is that we should be humble and not pretend that we invented the wheel. Therefore, we should refer to the original works and—particularly for those that do not agree with our opinions—try to change their statements as little as possible so that readers can make a fair judgment about which ideas they agree with. It has occurred too often to me already, and probably to many readers of this book, to see ideas wrongly presented in works by others, either when they agreed or disagreed with them, so I want to make sure I avoid that. Accordingly, when I cite a certain work, I will often use parts of the text written by the original authors. I am aware that in such textbooks one often tends to be more general and to not cite too often directly other works. However, this is against my philosophy, and I think readers will understand, and hopefully appreciate, it as honesty. Having said that, I am fully aware that I could have read more books and papers and added more references to the ones I included in this book because my tendency would naturally be to do so. But because I intend to have the Organic Nonoptimal Constrained Evolution (ONCE) idea proposed in this book read and hopefully discussed by a wide audience, I made the difficult decision to leave out some references. I also tried to reduce the jargon. However, I am conscious that I surely was not completely successful in doing so and that at least some parts of the book will still seem very technical, some sentences too long, and some historical comments unnecessary.

My main aim for this book is to try to provide an integrative, unifying vision about evolution and to help bridge the gap between various theories and lines of thinking that were presented for a long time as if they were conflicting with each other or were even irreconcilable. Thence my emphasis on bringing older ideas to modern discussions. Regarding the recent literature, I combine, for instance, some of the major points defended by the Extended Evolutionary Synthesis with some aspects that are, in my view, not emphasized enough in the books and papers about that synthesis. For example, the fact that the central active players in evolution are in general the organisms themselves and in particular their behavioral shifts and persistence was argued several decades ago by authors such as Baldwin and Piaget but neglected by most current researchers. Or the fact that eco-morphological mismatches are much more frequent than normally assumed. These mismatches are, for instance, related to the highly constrained character of organic evolution including the strong developmental constraints recognized in the Extended Evolutionary Synthesis. However, they are also associated with factors less emphasized in this synthesis such as the importance of behavioral persistence, which can dramatically limit the occurrence of new behavioral shifts and thus the responses to environmental changes often leading to evolutionary dead-ends and eventually to extinctions. In fact, one of the founding fathers of ethology, Niko Tinbergen (1953: 3), has emphasized how even behavior, which is the key driver of evolution according to ONCE and is often seen as highly flexible, is in fact also highly constrained: "Behavior shows wonderful adaptations, but also astonishing limitations."

Furthermore, ONCE also stresses another crucial point: the N of "Nonoptimal" which could also be the N of "Nonstruggling" because it refers to the fact that evolution is not necessarily a process where organisms are engaged in an incessant, suffocating struggle. Under ONCE, evolution is not simply a desperate, savage competition in which only organisms that have an optimal or almost optimal "match" with the environment that they inhabit can become "winners" and, as the famous Abba song says, "take it all." That is, within ONCE life is not unavoidably seen as a struggle 100% of the time for 100% of the organisms in 100% of their developmental stages. Life is more diverse and fascinating than that. As recently noted by Gailer and colleagues (2016), more and more studies are emphasizing the large plasticity between the so-called "optimal"morphology of a structure and the potential function of that structure, underscoring the need to appreciate apparently "maladaptive" structures in biological evolution as nevertheless effective func-tioning units. That is, such structures and the function they perform are "good" enough to allow the organisms displaying them to survive and reproduce, in the nonstruggling view of life defended in ONCE. As long as there is enough time and energy in this planet, there will be behavioral diversity and variation for mistakes, for trial and error, for neutral behaviors, and even for maladaptive behaviors on some occasions, i.e., for both etho-ecological and eco-morphological mismatches and for non-optimality.

Evolution in reality is generally made of mistakes, mismatches, and trial-and-error situations, which lead to new behaviors and that differentiate life and its complexity from the more deterministic existence of inanimate objects. In my opinion, the notion of a "struggle for life" has been blindly accepted for too long. It was mainly fueled by studies that were highly biased, *a priori*, to support this view, and the adaptationist program, e.g., the "just-so stories" mentioned by Stephen Jay Gould. Unfortunately, this view of evolution has led to many wrong ideas in evolutionary biology that unfortunately lead to catastrophic events in human history and that have obstructed, and continue to obstruct, a more comprehensive view of the diversity and complexity of life. An example is extreme adaptationism, which has fortunately been challenged in modern fields such as Evo-Devo but continues to prevail in areas such as evolutionary psychology and behavioral ecology. Currently it is even rising to new levels in recent fields such as evolutionary medicine, a field I particularly admire for its good intentions, but one that has also taken some problematic paths in my opinion.

As an aside, it was mainly a coincidence that the words I choose to express my view of evolution combined to form an acronym that corresponds to the title of one of my favorite movies. In the beautiful and Oscar-winning film *Once* (2007), by director John Carney, the characters played by actors Glen Hansard and Markéta Irglová spend much of their time in both a non-optimal and non-struggling existence, mainly singing and playing deep and powerful songs and being highly constrained by their past. Ultimately, they make a behavioral choice that is surely not the one that most spectators wanted to see them making, what does not mean it was the *wrong* decision, because the diversity, complexity, randomness, and shifts of life can always surprise us. For instance, years after seeing the movie, my girlfriend and now wife, the astonishingly beautiful and intellectually complex Alejandra Hurtado de Mendoza Casaus, with whom I originally saw the film, told me that she chose the main song of the movie for our wedding ceremony. And, years later, here it is, *once* again, now as the subject of the most personal book I have written so far. Because the dedication and cover of the book is dedicated to Tots for obvious reasons related to the main subject of the book, I therefore take the occasion here to thank Alejandra, the "once" of my life.

Washington, DC, USA   Rui Diogo

Acknowledgements

Apart from the hundreds of colleagues, students, and friends with whom I had the privilege to collaborate with and/or discuss the numerous topics covered in this book, from broader philosophical questions to specific details about the number of vertebrae in a particular marsupial species, I want to particularly thank the following individuals: Julia Molnar, for doing the very first detailed review of the whole book; Janine M. Ziermann because without her I would probably never have the time to write this book because she literally took care of all things when she was my postdoc and she continues to be a close colleague and friend and extremely helpful even after that; and Alessandro Minelli, Eric Parmentier, Virginia Abdala, Marcelo Sanchez, and Diego Rasskin-Gutman for reading previous drafts of the book and making pertinent and honest comments about it.

I further want to acknowledge, and also dedicate this book to, the advisors of my two doctoral degrees—Michel Chardon and Bernard Wood—who are completely different in many aspects but similar in one crucial one: They are open-minded and always willing to pursue the big questions and, above all, they promote the same type of behavior in their students, which is a true case of behavioral persistence. Moreover, Michel Chardon did a detailed review of a previous draft of the book and, as always, greatly contributed to make the book better. Last, I want to thank my parents, Valter Martins Diogo and Maria de Fatima Boliqueime, because although this might sound cliché, it is a self-evident and powerful, reality: Without them, I would not be here, and this book would not have been written.

Contents

Boxes

Chapter 1
Introduction to Organic Nonoptimal Constrained Evolution (ONCE) and Notes on Terminology

Never forget that only dead fish swim with the stream.
Thomas Malcolm Muggeridge (1903–1990)

A recent, well-written review by Morris on plasticity-mediated persistence [265: 8] defended the idea that developmental plasticity can allow epigenetic phenomena to "respond in remarkable ways to environmental input." One of the examples given was "the so-called 'two-legged goat effect', named by West-Eberhard after a goat that was forced to walk on two legs due to a congenital limb defect. I italicized the word "forced" to stress the fact that, although Morris is subscribing to a recent tendency within evolutionary developmental biology (Evo-Devo) toward a more organismal and less Neo-Darwinist and gene-centered view of evolution, he still used this word to describe the goat's behavior. Was the goat really forced to walk bipedally? Forced by whom? By a shepherd? By the "external environment"?

What actually happened is that the goat responded to a phenomenon that occurred in its body, i.e. the anomaly of its limbs, with one among many behavioral choices that it could take within the range of its developmental, behavioral, physiological, and/or anatomical plasticity. Of course, that behavioral choice also depended on the external environment in the sense that if, for example, the goat were in an aquatic environment, bipedal walking would not be a possible option. That is, in this case the external environment constrained the number of possible behavioral choices. However, it can hardly be said that the key active player in this example is an external environmental input: instead, it is the goat itself, in particular its behavioral choice. As can be easily seen with a simple Google Images search for "two-legged goat" or "two-legged dog", and as is shown in Fig. 1.1, there is a striking diversity of behavioral choices, postures, and specific types of locomotion that can be assumed by goats and dogs when confronted with a defect/injury of the forelimbs. The box below discusses and compares the chief features that define behaviors, behavioral responses *sensu* most authors, and behavioral choices *sensu* the present work, which are crucial to understand the ideas and empirical data presented/discussed in the present book.

© Springer International Publishing AG 2017
R. Diogo, *Evolution Driven by Organismal Behavior*,
DOI 10.1007/978-3-319-47581-3_1

Fig. 1.1 Selected sample within the strikingly diverse examples of behavioral choices made, and postures and specific types of locomotion assumed, by dogs when confronted with a defect/injury of the forelimbs (modified from public domain internet pages after a simple Google search for "two-legged dog" Google image search, namely from, https://www.gofundme.com/SHHRoo, https://i.vimeocdn.com/video/452201007_1280x720.jpg, https://www.youtube.com/watch?v=w1d Vlc_X9yM and http://4.bp.blogspot.com/_LBRxNEgFXaM/SU3djwGRlBI/AAAAAAAAAQ8/ y9Ch6Rz-mHU/s320/Chook+The+Two-Legged+Dog.jpg)

> **Box—Definitions: Behavior, Behavioral Responses, and Behavioral Choices**
> Unless stated otherwise, terms presented in this book will usually be related to their most commonly accepted, widespread definitions to avoid unnecessary confusion and excessive use of jargon. Accordingly, when I use the term "behavior" I refer to a very simple and broad definition often seen in the literature: behavior is a conscious or unconscious response of an organism to stimuli or inputs, i.e. it applies to all living organisms without exception.

Accordingly, the term "behavioral choice" refers to cases in which at least more than one potential choice is possible: the organism chooses, and then has the drive to undertake, one of the possible choices. This is considered a behavioral choice no matter—again—if the choice and/or drive to undertake it are conscious or not. It is likely that behavioral choices in organisms such as bacteria are often unconscious, whereas in organisms such as humans, chimpanzees and elephants they are usually conscious.

As will be explained and discussed in detail throughout this book, under the definition adopted here, there are three key items for a certain behavior to be considered a behavioral choice. The first is that more than one behavior/outcome is possible, i.e. it is not merely like a rock that falls down from a cliff by "moving" due to the force of gravity, in which the outcome is always the same: the rock will always move downward toward the center of the earth. If we now think about a bird in the air, it experiences the effect of the same force of gravity, but there are many possible outcomes, which are thus behavioral choices. The bird can let itself passively fall downward toward the center of the earth. However, it can instead fly to counterbalance the force of gravity by staying at approximately the same altitude, or it may decide to fly even higher to a greater altitude, and so on. This leads us to the second key item: drive. To undertake a behavioral choice, the bird must have the drive to do so, i.e. the bird is an active agent precisely because there is more than one option and possible outcome, and the bird thus needs to choose and undertake just one of them. Therefore, contrary to inanimate objects, living beings are not automata: rather they are active players. This in turn leads us to the third key item: behavioral choices are always undertaken by the organism as a whole: they are organismal behaviors and cannot refer to the behavior of just parts of the whole, such as for example atoms, cells or tissues.

Therefore, an important point is that behavioral choices, as defined here, are just a subset of the "behavioral responses" *sensu* Sultan [350: 51], which do not need to meet all of the three defining key factors listed previously. In fact, they may even lack all of them at the same time. As explained by Sultan, "a quantitative distinction is sometimes drawn between 'intentional' behavioral responses, which are mediated by a central nervous system, and the kind of growth and movement changes expressed by plants and other non-animal taxa in response to specific environments … this distinction has been challenged by several plant biologists, who have argued that plant evo-devo responses to environmental information constitute behavior and, indeed, intelligence." She then states that "in a broader framework, all types of environmentally mediated phenotypic expression can be viewed as cue and response systems."

However, in my opinion, by following this latter statement, Sultan is missing—as are most Evo-Devoists and biologists in general—the crucial distinction between two types of phenomena with completely different biological and evolutionary meanings. One concerns behavioral responses that are merely physiological and thus sometimes nearly "automatic" responses similar to, e.g. the "response" of my computer when I click the "delete key" while writing this book. The computer does respond to this stimuli (i.e. pressing this specific key) by displaying a "behavior" according to the widely used definition given just above. However, two key defining features of behavioral choices are lacking. First, the computer always responds in the same way to this stimulus: there is only one single possible outcome, which is thus totally predictable. The computer will not surprise us with a different outcome because it lacks the second key item: it has no drive to select a different outcome because it has no drive at all. In a similar way, when I am outside and the temperature reaches 40 °C, my skin will respond always in the same predictable way: I will sweat. That is a behavioral response *sensu* Sultan, but is merely a physiological response that has nothing to do with an organismal behavioral choice because it lacks all three key defining items: there is only one outcome, and it is not a choice, nor does it involve a drive of the organism as a whole because I do not choose—and surely do not have the drive—to sweat. Actually my drive is to avoid sweating, and that is why I normally do not stay outside when the temperature is ≥ 40 °C.

Therefore, the main dichotomy should not be between having "a central nervous system" as animals do versus not having one as in plants. Instead, it is having the capacity to make behavioral choices and the drive to undertake them, and for this organisms do not need to have a central nervous system because all living beings, including plants and bacteria, clearly display behavioral choices, as I will explain in detail throughout this book. A crucial concept—from the field of systems biology—is thus emergence, in which the organism can display a behavioral choice as a single unit, no matter whether or not it has a central nervous system or any type of consciousness. The dichotomy between organismal behavioral choices versus other types of behaviors can thus match the one Sultan referred to, about having or not having "intentionality". However, this only applies if the term refers to the drive that the whole organisms must undertake to carry out certain behavioral choices, and not necessarily to consciousness nor to any teleological concept related to "evolutionary purpose or goal" (see later text).

Numerous examples of the "behavioral responses" given in Sultan's [350] excellent book and many other publications cannot thus be considered at all to be behavioral choices sensu the present book. Nor are those examples related to any type of teleological "purpose", "intentionality" or "goal", as it is often suggested in the literature. For instance, if a certain lake starts having too many predators, and various salamanders living in that lake start meta-morphosing as a "response" to this increase, this is not due to a behavioral choice of the salamanders nor to a drive of the salamanders to do so, and surely not due to any "final goal" of the salamanders. It is instead an epi-genetic response that leads to a single possible outcome. As seen in many laboratory experiments, by changing and combining factors—such as the total volume and composition of the water, the number of predators and salamanders, and so on—one can easily predict whether the members of a population of a known species of salamanders will tend to metamorphose or not. The salamanders cannot do anything about it, just as I cannot stop sweating when I am in a sauna, because this is merely a physiological response beyond my control, which I can only avoid by undertaking, for instance, the behavioral choice of leaving the sauna. Accordingly, many other examples of behavioral response provided by Sultan do conform to my definition of behavioral choices, e.g. some of the phenomena she designates as "habitat choice behaviors" in which organisms seek, or avoid, particular climatic conditions. Examples include the behavioral choice of anteaters to switch the foraging activities to nighttime in extremely hot weather, of many reptiles to bask in the sun to increase body temperature during cold weather, and of penguins, mice, bats, marmots, and pigs to aggregate in order to decrease their collective surface-to-volume ratio in cold conditions.

In summary, as Lindholm [232] put it, behavioral choices cannot be reduced to genetics—or, I would add, to mere automatic, physiological, or localized/regional epigenetic reactions to external stimuli or other factors—because this requires a subject to take choices and have the drive to undertake them, which is the whole organism. This capacity and drive to undertake behavioral choices obviously depends on intrinsic genetic or genomic, and epigenetic (e.g. hormonal or physiological), features linked with external factors. However, as noted above the capacity is ultimately mainly related to a phenomenon that is now becoming more and more prominent in biology, particularly due to the rise of systems biology: emergence. That is, a strikingly high number of complex factors, both intrinsic and extrinsic, are combined in a way that an overall outcome that is more than just the sum of the part emerges: the capacity to take a behavioral choice and having the drive to undertake it. Contrary to mechanistic and atomistic views that have prevailed for a long time in the history of biology (see later text), this capacity does not apply to any of the organismal subunits or regional parts/organs. Individual atoms or electrons do not walk bipedally as we do, nor can they choose to do so. This capacity only applies to the whole organism, thence the term "organismal behavior".

The use of the term "forced" in the example of the two-legged goat is even more striking considering that in the very same article, Morris [265] supported Baldwin's idea of organic selection, in which behavioral choices, followed by behavioral persistence across generations by way of social heredity, play a crucial role in evolution. This apparent paradox therefore has nothing to do with a bias of Morris himself. The paradox is instead mainly explained by a historical bias that is ultimately related to a major event in the history of Western science: the so-called "scientific revolution" that occurred in the sixteenth and seventeenth centuries as explained in *Beyond Mechanism—Putting Life Back into Biology* [182].

Box—Definitions: Behavior, Behavioral Responses, and Behavioral Choices
Behavioral persistence refers to cases where the generations that follow those
in which a certain behavior was originally acquired continue to respond to the
same stimuli or input in a similar way, mainly due to social heredity (e.g.
imitation, learning, and/or teaching), as explained below. Regarding epigenetics, I use the term in the *sensu lato* and follow Sultan's [350: 9] statement
that "epigenetics effects at the molecular level can be defined as biomechanical
mechanisms that shape patterns of gene expression in the absence of any
change in nucleotide sequence". Therefore, DNA methylation or histone
modification are only specific examples among many other epigenetic phenomena that can be listed, including, e.g. those associated with endocrine
activity, as beautifully and extensively reported by Matsuda [245]. An updated
list of such other epigenetic phenomena as well as recent experimental and
medical studies on the subject—plus a discussion of their evolutionary significance including their links with some (but not all [see Chap. 5]) of
Lamarck's ideas—is given in Sultan's [350] book *Organisms & Environment*.

**Box—History: The Scientific Revolution and the Historical Context of
Darwin's Ideas**
The scientific revolution "involved many major changes, including
Copernicus's putting the Sun at the center of the universe, Kepler's work on
the planets, Galileo's and Descartes' articulation of mechanics, and finally
Newton's great synthesis" [320: 409]. As noted by Ruse, contrary to Greeks
such as Plato and Aristotle, who "subscribed to an organic view of the
world", "after the Scientific Revolution people subscribed to a mechanic view
of the world; Robert Boyle saw things clearly—from now on we work in
terms of machines, of artifacts: organicists like Aristotle who see living forces
directing nature are just plain wrong." In particular, "the concepts of 'function', 'doing', 'purpose' and 'agency' in biology, totally absent in physics
where only 'happenings' occur" have become mainly muted in standard
biology by a concept expressed by Jacques Monod [207: 4]. This concept
explains why many scientists could not resist Darwin's theory and analogies
in the nineteenth and twentieth centuries: "when phenotypic variations of a
species are interpreted as a state and the selective pressure as a force, exerted
by external conditions, a seductive analogy to mechanical operations is
obtained" [130: 94–95].

It is surely not a coincidence that Darwin [67] explicitly referred to gravity, and
thus to Newton's mechanicism, in the last sentence of his most influential book:
"there is grandeur in this view of life, with its several powers, having been originally breathed into a few forms or into one; and that, whilst this planet has gone

1 Introduction to Organic Nonoptimal Constrained Evolution …

cycling on according to the fixed law of gravity, from so simple a beginning endless forms most beautiful and most wonderful have been, and are being, evolved." As Hoffmeyer [187: 148–152] put it, "Darwin created a perfectly externalist theory, a theory that seeks to explain the internal properties of organisms, their adaptations, exclusively in terms of properties of their external environments, natural selection pressures." Darwin was a passionate naturalist who observed nature in incredible detail; but because of the historical constraints, particularly the scientific context of his epoch, he tried above all to connect evolution with external forces. However, as noted by Hoffmeyer, although Darwin "managed to construct an externalist explanation for evolution, he was not a fundamentalist in his externalism as were his followers (the Neo-Darwinists) in the twentieth century, who thought they could get rid of organismic agency by enthroning the gene and seeing organisms as passive derivatives of genotypes." In fact, Darwin did made a crucial distinction between "external" natural selection and selection related to organismal behavioral choices, which broadly corresponds to Baldwin's organic selection (see Chap. 4), but this was a key point that Neo-Darwinists mainly ignored. Such a Neo-Darwinist view of evolution continues to be followed by many—probably most—biologists as exemplified by Richard Dawkins's famous quote that organisms are "no more and no less than survival machines" [319: 414]. Even authors such as Edward Wilson— who described insect societies as "emergent" forms of social order that arise through the collective "decision making" of individual insects—at times described the insect colony as a "growth-maximizing machine" composed of "cellular auto-mata" whose operations can be described by the language of physical and computer science [181: 236].

Box—Definitions: Adaptation, Adaptationism, Externalism, Internalism, and Internal Selection

Regarding the definitions of these terms, I will mainly follow the more fluid, engaging style of authors such as Olson [278], focusing more on the ideas and, in particular, on how authors with different views can defend them using similar definitions and data. I will thus begin by using Olson's [278: 278] very simple definitions: adaptation as "the process by which form comes to reflect function as the result of the action of natural selection"; adaptationism as "thinking of adaptation as the cause of the form-function fit"; internalism as "the view that the dynamics of the developmental system so channel developmental potential that they are the primary directors of the evolutionary process"; and externalism as "the view that developmental potential is so vast that the outstanding directors of evolution are factors external to the organism". Then, throughout the book, I will show how some of these definitions, or the main ideas on which they are based, are sometimes too simple or even incorrect.

For instance, one of the key ideas defended by Olson, i.e. that cases of developmental internal selection provide support for an adaptationist view of evolution, does not fit in the context of the definitions he provided. The terms "adaptation" and "adaptationism" continue to be mainly associated with the Neo-Darwinian view of evolution, which focused in great part on the fit between adult form and function. As recognized by Olson [278: 283], "one of the most notorious aspects of the modern synthesis has been its elision of development, treating it as a trivial, more or less deterministic black box between the genome and the phenotype." Olson also recognized that within this context, adaptationism was mainly a synonym of externalism with natural selection, in particular the external environment, playing a central role in morphological macroevolution, although Neo-Darwinists also stressed the importance of other factors in evolution, e.g. genetic drift and gene flow. However, as stated in Futuyma's book *Evolution* [136], natural selection has historically been seen as the only mechanism known to cause the evolution of adaptations, i.e. of the processes in which the members of a population become "better" suited to some features of their environment through changes in characteristics that affect their survival or reproduction in that environment.

Ecomorphologists have been, and in a way many continue to be (e.g. [230]), inspired by such views because they are interested in the links between morphological adaptations seen in adults and the ecology/external environment they occupy (a succinct account of the history of adaptationism is given in the first chapter of Wagner's 2014 book *Homology, Genes, and Evolutionary Innovation*). For example, ecomorphologists have stated that a numerical increase in cervical vertebrae is an adaptation to herbivory in theropod dinosaurs [408]. Such ideas may prove to be correct. However, it is too much of a stretch to argue, as Olson [278] did, that Evo-Devo evidence that internal factors play a crucial role in morphological evolution is leading to a renaissance in adaptationism because in many cases unoccupied adult morphospaces can actually be produced early in ontogeny but are then eliminated by internal selection. He correctly cites, as an example of internal selection, the occurrence of eight cervical vertebrae in fetuses/babies/children that is often associated with deleterious conditions that leads to death at an early age [138]. However, this example of internal selection supports the original point made by Galis, which is precisely the opposite of the point that Olson is trying to make. This example emphasizes that internal selection is a subset of internal factors that limit the adult morphospace and that are essentially independent of the external environment, therefore going against an adaptationist/externalist view of evolution. Humans at early developmental stages do not die because of the physical presence of an eighth cervical vertebra that is negatively selected for by the external environment. Instead, having an eighth vertebra is just a byproduct of the collapse of the whole internal developmental homeostasis at a very sensitive ontogenetic period, the so-called "phylotypic stage" ([138]; see Chap. 7). One should thus not confuse the works and ideas of Galis, which inspired the new Evo-Devo subfield of Evolutionary Developmental Pathology and Anthropology [106] with the extreme panselectionist form of adaptationism that is followed by many researchers within the rising field of Evolutionary Medicine [363], to which I surely do not subscribe.

In fact, as pointed out by Schwenk and Wagner [330: 57], it is external (natural) selection that "changes with the external environment, whereas internal selection remains essentially constant because it travels with the organism" and is "imposed by intrinsic, organismal, functional integration, even as the external environment changes." Or, as put by Wagner and Schwenk [387: 157], internal selection is "independent of external, environmental selection pressures (because the limitations are imposed by intrinsic attributes of the organism); as such, it is the internal coherence and functionality of the system, as a whole, that imposes its own 'internal selection' on individual characters, determining which character variants are viable." Over millions of years, almost all adult mammals in many different external environments and geographical locations had only seven cervical vertebrae because of such internal developmental factors (i.e. constraints [see later text]).

Of course, Darwin's "external" natural selection exerted over a taxon involves not only purely abiotic factors (e.g. a meteorite affecting our planet, or, an "ice age") but also biotic factors, which are in turn often related to behavioral choices made by organisms of other taxa. For instance, a major aspect of the "external" natural selection—which corresponds to natural selection *sensu* Baldwin and *sensu* the notion of Organic Nonoptimal Constrained Evolution (ONCE) presented in this book (see Fig. 1.2)—exerted on salmon in rivers in Alaska is related to the behavioral choices of bears to catch and eat them. However, the point of using different terms, as Baldwin did and as done in this book, is precisely to stress the fact that Darwin's natural (external) selection is the sum of a huge network of abiotic and biotic environmental factors and that those biotic factors are often related to the organic selection, and thus the behavioral choices, of other organisms, therefore emphasizing the crucial role of organismal behavior in evolution. Organisms not only make their own behavioral choices and thus help to construct their own niches, but they also interfere with the selective pressures that will be felt by other organisms and therefore influence the niches that those other organisms help to build. It is a hugely complex and dynamic network that is far from a mechanistic view of evolution and from the idea that evolution is mainly related to "active external forces applied to passive organisms". Therefore, in this case, if the subjects of our discussion are the salmon, the use of the term "organic selection" by Baldwin and in ONCE is related to the behavioral choices of the salmon themselves, e.g. to go back to the place where they were born, or to the preferences of the females for certain specific features of the males. Then, within the context of the niche/way of life chosen by salmon, natural (external) selection *sensu* Baldwin and ONCE refers to factors that are external to salmon. These include biotic factors such as predation by bears and abiotic factors such as the physical water currents that salmon must face when going back up the river to the place where they were born.

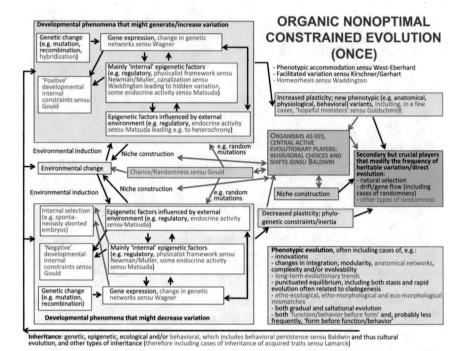

Fig. 1.2 The structure of the idea of Organic Nonoptimal Constrained Evolution (ONCE) presented in this book. Terms and arrows shown in red are those emphasized by ONCE but not in the Extended Evolutionary Synthesis scheme shown in Fig. 1.3. Following the style of that Extended Evolutionary Synthesis scheme shown in Fig. 1.3, arrows represent causal influences (see caption of Fig. 1.3 below for more details). Within the context of the present book, a particularly crucial differences between ONCE and EES, emphasized in a central position of the scheme of this Fig. 1.2 with larger fonts, is the fact that in ONCE organisms, and in particular their behavior (organismal behavior) are seen as the key active players of biological evolution, with Darwinian (external) natural selection playing mainly a secondary—but still crucial—evolutionary role. This "organic" (the "O" of ONCE; based on Baldwin's notion of "Organic selection") view of evolution, together with the "nonoptimal", "nonstruggling" view of evolution also defended in ONCE (its "N"), enables a natural, rational, non-teleological and non-vitalistic explanation for the common occurrence of long-term evolutionary trends, as well as of etho-ecological, etho-morphological and eco-morphological mismatches (for more details, see main text)

Within the context of Baldwin's organic selection and regarding the part of ONCE that refers to that concept (Fig. 1.2), one can thus say that organic selection, i.e. the behavioral shift in salmon to return to their place of birth and their subsequent behavioral persistence to continue doing so for many generations, was an active key player in the evolutionary history of salmon. Natural (external) selection was a secondary, but also hugely important, player because within the niche/way of life that salmon help to construct, advantageous phenotypic features—including morpho-logical (e.g. having fins that allow them to move faster) and/or physiological (e.g. having muscles that can contract faster/more powerfully) ones—were then selected by such an external selection. Of course, regarding the bears, organic selection refers to their behavioral choice to go to the rivers to catch and eat salmon, and natural (external) selection refers to abiotic and biotic factors external to the bears. These may include river currents that might make it easier or more difficult to catch the fish, the way the fish live and move, and so on. Cases of so-called "evolutionary arms races", such as the interaction between bears and salmon, thus involve a very complex interplay between the organic selection of each group of organisms and natural (external) selection. That is, a mix exists between the drive of each group separately to display a certain behavior or to, e.g. select mates in a certain way, and behavioral responses to the behavioral choices of the members of the other group, and/or to other external biotic, and/or to abiotic environmental selective factors.

It is mainly because the mechanistic view has been historically defended by many Neo-Darwinists, as well as its historical background, that terms such as "conscience" and "intelligence"—as well as related discussions on whether or not only humans display a "true" culture—are so controversial and often turn into endless philosophical and/or nomenclatural debates. Even the use of the term "imitation", which at first glance would seem more consensual, has been a major subject of controversy because some researchers cannot admit that "imitation" performed by at least some non-human organisms is really similar to imitation within humans [75]. These controversies explain why even authors such as Morris [265] might feel uneasy referring to or using such terms in the context of non-human animals and thus often opt to simply use more "neutral" terms such as "the goat was forced". However, by using these terms, one risks perpetuating the idea that the role played by non-human animals in their own evolution, and even in their day-a-day life, is necessarily passive. That is, that organisms display certain behaviors only because their genes (e.g. by "random mutation"), the external environment (e.g. "natural selection"), and/or humans (e.g. "artificial selection") forced them to do so. In this view, the organisms themselves seem to have no drive, no options, and no active role: they seem to merely exist, to merely survive. This idea is in fact stressed in the emphasis put on "survival" in the definition of natural selection as "the differential survival and reproduction of individuals due to the differences in phenotype". The idea that organisms as a whole play merely a passive and insignificant role in evolution is even more emphasized in the Neo-Darwinist definition of evolution as "changes in allele frequencies within populations." The specific view that genes are "selfish" replicators, that phenotypes are perishable vehicles, and that life is optimized for genes rather than for organisms, has therefore emerged as particularly persuasive in large subsets of the academic community, as well as the general public and press, since the 1970s [232].

This discussion leads us to the question: Why is this book needed? According to Downes [115: 41], one should not need to invoke explanations other than Neo-Darwinism unless one needs to explain phenomena "that appear to go against what one would expect from an evolutionary (Neo-Darwinist) standpoint." That is why I am writing this book: because there are two major facts, among many others that I will discuss in this work, that go against what would be expected from a purely Neo-Darwinian view of evolution. One has been discussed in several books and specialized papers without any consensus being reached so far: the occurrence of long-term macroevolutionary trends, which is related to the teleological concepts of "purpose" or "progress" in evolution (see box below), and to the difficulty of explaining the origin of morphological novelties. The other has been less discussed because it has only come to be shown to occur consistently in more recent phylogenetic ecomorphological studies: the common occurrence of etho-ecological, eco-morphological, and etho-morphological mismatches. That is, in the last few years, numerous papers—most of them written by ecomorphologists to identify/confirm correlations between ecology and anatomy—consistently reveal eco-morphological mismatches in which form is, for instance, much more strongly related to the phylogeny of a group of organisms than to the current ecological habitats inhabiting by them. The frequent occurrence of such mismatches was not at all expected in the light of the works of Neo-Darwinists, particularly those subscribing to an adaptationist framework—and, for that matter, even of Darwin himself, who for instance stressed how the morphology of the Galapagos finches seems to be beautifully optimized for the specific habitats in which they live.

> **Box—History: Aristotle, Scala Naturae, Religion, Design, Purpose, Teleology, and Vitalism**
> Subjects such as the passive versus active role of organisms and related topics as the notion of evolutionary trends, the idea that complexity supposedly increases during evolution, and the form versus function debate, have been crucial within the history of biology since Aristotle and particularly of evolutionary biology since Darwin, Wallace, and others [109]. These subjects are also related to the old concept of scala naturae (ladder of nature from "lower forms" to humans, which supposedly represent the culmination point of a "progression" toward perfection) and to associated teleological topics such as the notion of "design" or "purpose" in evolution [236]. Specifically, such teleological notions are linked to the question "why are organisms constructed so well to perform their functions?" [361: 193].
>
> Reiss' [305] *Not by Design* provides a detailed and well-documented summary of the history of teleological reasoning from the Greeks to modern times. I thus refer the readers to that book for more details on this topic and strongly recommend the book to anyone interested in evolutionary biology in general. According to McShea [255: 665), a major reason why scientists have been so interested in such topics for millennia—which in turn makes discussions on these issues so difficult and often contentious—is that "there has always been an aura of mystery, of magic, around such systems on account of their seeming future directedness." McShea further notes that "the three standard terms of discourse (teleology, goal-directedness, purpose) all imply a future object or event (a telos, a goal, an achieved purpose) that is in some sense explanatory of present behavior." This last sentence also emphasizes the idea that even among those authors who defend the existence of evolutionary trends related/leading to new behaviors/phenotypes, these trends were historically usually seen as driven by other forces, e.g. either vitalistic, or created by an intelligent designer, or selected by the external environment. Within all of these diverse ways of thinking, organisms were once again

mainly seen as being passive within the larger scheme of things created by such powerful forces.

Examples of the renewed interest in these issues include the publication of several books about them in the last 25 years (see also, e.g. Johnson et al. 2002; [268, 282, 310]). These include, among many others, *Randomness in Evolution* [45], *Evolution Without Darwinism* [49], *Forms of Becoming* [259], *Modular Evolution—How Natural Selection Produces Biological Complexity* [369], *Not by design—Retiring Darwin's Watchmaker* [305], *Niche Construction – The Neglected Process In Evolution* [276], *Darwin and Design—Does Evolution Have a Purpose?* (319], *Arrival of the Fittest: Solving Evolution's Greatest Puzzle* [385], and *The Music of Life: Biology Beyond the Genome* [272].

A central tenet of ONCE is actually that, contra the notion of "design", in most cases organisms are not so well constructed to do what they do/how they live in their current habitats. Therefore, ONCE also has no connection with a mysterious vitalistic force within living tissues driving organisms in a single direction, or to "perfection", as proposed by some defenders of orthogenesis in the past (see Chap. 4). Instead, a certain behavior choice by an organism/a group of organisms is simply one among many possible choices as emphasized previously. For instance, some organisms are positively phototactic (i.e. tend to move toward the light), whereas others are negatively phototactic because, among the many options available and possible within the context of their own plasticity and constraints, the choices they made were viable in that specific case, time, geographic location, and habitat where they took place. Therefore, there is nothing vitalistic or deterministic here. Instead, ONCE assumes precisely the opposite. That is, ONCE does not hold if there were a simple, unidirectional, constant force guided by simple natural laws, such as gravity making a stone rolling downhill, nor by an internal program or genome into which a deterministic goal of the organism has been "coded" as in a homing torpedo [255]. As noted by McShea [255], apart from plasticity and behavioral persistence, theories such as ONCE require partial independence between the organism and the various forces being exerted on it because this is also part of what makes behavioral plasticity, choices/shifts, and persistence possible. Partial independence gives the entity the capacity to make errors, or to be deflected in arbitrary ways, and to respond with corrections unlike a stone rolling downhill. For McShea, it is the constant error-or-deflection followed by correction that generates their signature, and apparently teleological, persistent behavior.

It is particularly interesting to note that even Darwin's finches, which are often used as a landmark case study for Darwinists and Neo-Darwinists and for the concept of allopatric speciation—i.e. random changes in allelic frequencies allow fitness to vary, which in combination with geographic isolation eventually lead to new species—actually stress the crucial role of behavioral choices and shifts in evolution. This is because the finches and other organisms that colonized the

Galapagos Islands were not merely passive players in this story: if that were the case, one would see similar patterns of speciation and/or morphological changes related to adaptations to the habitats of each island in other taxa as pointed out by Lindholm [232: 449–450]. However, the total numbers of bird species, excluding seabirds, on the islands amounts to 30, and of these 14 are finches; among the others, only 1 group shows any sign of radiation. In turn, rodents show some degree of speciation, whereas bats do not, and among the 1500 insect species on the Galapagos Islands there are hardly any new species. Thus, as stressed by Lindholm, "the impressive radiation in Darwin's finches is clearly not part of a general pattern; all colonists found similar unoccupied niches at the archipelago, but their evolutionary (and behavioral) response differed." In fact, the Grants' three-decade observation of the finches has clearly shown, for instance, cases of immigrants from a particular island that seemingly scanned very carefully the novel habitats of the new island for certain features preferred within the context of their previous behavioral persistence, and thereafter accordingly chose, based on the information obtained from that scanning, to breed on that island or to move on.

In this book, I argue that both problems mentioned previously (i.e. mismatches and long-term evolutionary trends) can be explained with the idea of ONCE summarized in Fig. 1.2. I should point out straight away that I made a peculiar behavioral choice for the book. The major tenets/bullet points of ONCE are schematized in Fig. 1.2 and mentioned in this chapter to help guide the readers through this and the next chapters of the book. However, I will intentionally only fully develop the details of ONCE throughout those later chapters and will only fully integrate and summarize these details in Chaps. 9 and 10, when I will put together the topics presented/discussed in the first eight chapters. I did this because I am opposed to putting the cart before the horse. First, I want to present in detail the empirical and theoretical foundations that led me to think about ONCE so that the readers can follow me in this journey and know the line of reasoning I took each step of the way.

Therefore, to let the readers continue to experience my process of writing this book, I will proceed to the next step of my reasoning, which is this: if natural selection by the external environment were almost always the key force in evolution, one would not expect the frequent occurrence of etho-ecological mismatches and/or ecomorphological mismatches. In contrast, the more important the role played by internal factors—including both internal constraints and selection (see Chap. 7)—the more one would expect such mismatches to eventually occur. The idea of ONCE thus goes against not only the Neo-Darwinist view of evolution in general but also against many of the specific ideas traditionally followed within the adaptationist framework including works by functional morphologists and ecomorphologists that try to establish a "functional utility" for every single morphological and behavioral trait of every single organism. Humans are still bipeds in every type of external environment in which we live. Even when we were on the moon, we walked bipedally as attested by the famous sentence "one small step for man ...". An engineer might prove that it would be wiser to, for example, move quadrupedally on the ice of Antarctica, or in another non-bipedal way in the moon, but we would still walk on our two hindlimbs. This type of behavioral choice and persistence helps to explain why we see not only etho-ecological but also eco-morphological mismatches so frequently, in which phylogeny is a much better predictor of form than is ecology, as noted previously (see Chap. 6).

For now, I will just provide an illustrative empirical experimental example, reviewed by De Wall [75], of how behavioral choices and persistence might often lead to mismatches. Researchers gave vervet monkeys in a game reserve open plastic boxes with maize corn. There were always two boxes with two colors of corn, blue and pink: one color was good to eat, whereas the other was laced with aloe, thus making it repulsive for the monkeys. Depending on which color corn was palatable, and which was not, some groups choose to eat blue maize, and others chose to eat pink maize (behavioral choices). Researchers then removed the distasteful treatment and waited for infants to be born and new males to emigrate from neighboring areas to the group. All adults continued eating their acquired preference (behavioral persistence) and thus never discovered the improved taste of the alternative color, i.e. they did not adapt to the new ecological conditions (etho-ecological mismatch). With one exception—an infant whose mother was so low in rank that he could hardly eat—all 27 newborn infants learned to eat only the locally preferred food, not touching the other color, even though it was freely available and just as good as the other. Male immigrants, too, ended up adopting the local color even if they arrived from groups with the opposite preference.

This example is therefore also useful to show how adaptationist studies can be flawed: if an adaptationist researcher did not know what had happened and tried to study the current scenario, he/she would probably assume a priori adaptationist reasons for it and try as hard as possible to find the evolutionary advantage, for the monkeys, of not eating corn of a certain color. For instance, they might think that this happened because of a certain component of that corn or because of a very complex strategic setting by the monkeys in which it would be beneficial for them in the long run to eat a certain color of corn, and so on. However, there is no advantage at all, actually it is the contrary, because both colors are now palatable and thus a useful source of nutrient for the monkeys. The current scenario is instead simply the result of a mix between contingency (e.g. history including behavioral persistence and inertia [i.e. behavioral conformism *sensu* De Wall; 2016]) and chance in the sense that all began with an experimental idea, among many others that were possible to design and undertake, that was undertaken at a random point in time by a random member of a completely different species. The obsessive seeking for evolutionary advantages under the adaptationist paradigm is, ironically, clearly a *non-advantageous* scientific research methodology.

> **Box—Definitions: A Plethora of Definitions of Constraints in the Literature**
> Schlichting and Pigliucci [327: 166–168] made a meritorious attempt to "distill" the concept of constraints taking into account the plethora of ways in which other researchers have referred to them. I will thus refer to their work to show how many types of constraints can be distinguished in the literature

as well as how, in my opinion, some of the definitions proposed in the literature, including their own, suffer from both practical and theoretical limitations. They first define genetic/epigenetic constraints, which can result from the lack (or limitation) of appropriate genetic variability and/or from genetic linkage (e.g. alleles closer together on a chromosome being inherited together), genetic pleiotropy (e.g. a genotypic trait affecting more than one seemingly unrelated phenotypic features), genetic dominance (e.g. one allele being dominant and the other recessive), or genetic epistasis (e.g. an allele masks the visible outcome of/phenotype affected by another gene). Therefore, somewhat confusingly in my opinion, they define developmental constraints as a different type of constraint, although they recognize that this is a "mixed category than can include both selective pressures and genetic constraints." They thus define developmental constraints as "restrictions on the potential evolutionary trajectory due to the fact that the developmental system must maintain an internal coherence." Complicating things a little bit further, they add that "since the objective is optimization of the whole organism, this is an evolutionary force and not a constraint." Apart from the confusion generated by their use of these terms, by the explanation that they give, and by the adaptationist and teleological terms they use ("objective" of whom?), they cause even more difficulties by defining mechanical and functional constraints as a single type of constraint.

An example of functional constraints as they are often presented in the literature is the configuration of the abductor of the little finger. The only way to have an abductor of the fifth digit in a pentadactyl hand (i.e. a hand with five digits) is to have a muscle that inserts on that digit, in particular to its ulnar side, to generate an ulnar movement of the digit, i.e. an abduction [104]. This type of constraint is thus likely to be related to natural selection in the sense that if a disturbance of development results in a different configuration the animal would not be able to abduct its fifth digit; if such an abduction is crucial for its survival, the animal could eventually not be able to survive/reproduce. However, when Schlichting and Pigliucci [327: 166–168] state that "mechanical and functional constraints" are related to "physical laws", they further complicate the issue. One would think that "mechanical" constraints would refer mainly to developmental internal factors/constraints such as the physical limitations of how cells can aggregate, and so on, i.e. to Newman's/Müller's physicalist framework as noted previously. However, Schlichting and Pigliucci then state that "mechanical constraints, however, also arise due to selection and are thus a category of selective pressures." But the types of forces invoked in the physicalist framework are not so directly linked to internal selection. The physicalist properties limit the number of possible outcomes as well as the size of the occupied morphospace due to physical laws (e.g. related to gravity). It is not the case that an "error" causes a cell to fail to feel the force of gravity and then selection exerts selective pressure against it and the cell dies. As explained previously, internal selection instead refers to cases of such "errors", where there is, for example, a disturbance of development. A form that is normally achieved, e.g. a "normal" human face, is not achieved, e.g. resulting in a cleft palate, and the effects of that form, and of all the other pleiotropic effects related to/that have caused it, lead to the death of the organism even if the organism is inside the mother and not in direct contact with the external environment. Both the physicalist framework and internal selection can be included within the broader term "developmental constraints". However, in my opinion one cannot say that the former (e.g. physical laws that existed long before the origin of life on this planet) is a "special form of natural selection".

Apart from the selective constraints mentioned previously—which Schlichting and Pigliucci defined as possible regimes of selection, e.g. directional, disruptive, and stabilizing, which have a "negative" role in evolution—they also define ecological constraints, which result from a particular kind of selective pressure arising from the interactions of an organism with its abiotic or biotic environment. In my opinion, this is an example of the tendency of authors to overcomplicate concepts and/or create too many terms, which can confuse readers. That is precisely one of the major reasons why in this book I avoid focusing too much on theoretical definitions and why in general I prefer to focus on the ideas themselves, including the historical context in which they were developed, and on how these ideas are supported, or contradicted, by the actual data obtained in empirical studies.

Within ONCE, "developmental internal constraints" simply refers to internal factors that limit the regions of the morphospace occupied by the adult individuals of a population without the interference of internal selection (see previous boxes). Examples of such factors include physical forces as well as the effects and/or processes stressed by the physicalist framework *sensu* authors such as Newman and Müller, which include adhesion and differentiated adhesion of cells and self-assembly on anisotropic components (e.g. [270]) (Fig. 1.3). Genetic drift is a different phenomenon because it refers to the change in frequency of a gene variant within a population due to random sampling of organisms, i.e. it is not directly dependent on or directed by external factors. As noted previously, an entire book could be written about these definitions and how in some specific theoretical examples they can blur together, how alternative and more complex definitions could be given, and so on. An example of such a book, which includes Schwenk and Wagner's chapter cited just above, is *Keywords in Evolutionary Developmental Biology* (Hall and Olson 2003). In that book, the point is made that some other authors may have slightly different definitions of external versus internal environment, and that apart from internal factors, external factors can be extremely influential during early stages of development. Such factors range from differences in temperature or luminosity outside of the egg/inside the mother to whether a mother drinks alcohol or is killed by predators.

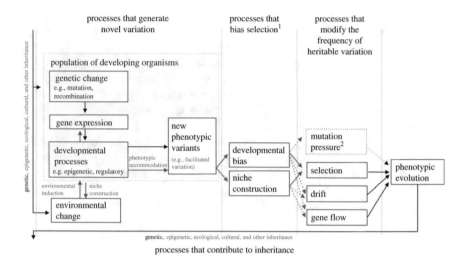

Fig. 1.3 The structure of the Extended Evolutionary Synthesis (EES) according to, and modified from, Laland et al. [218] with arrows representing causal influences and processes shown in red being those emphasized by the EES but not by a more traditional Neo-Darwinist perspective. EED includes as evolutionary causes processes that create new variants and bias selection, change the frequency of heritable variation, and contribute to inheritance. A diversity of ontogenetic processes (e.g. epigenetic effects, regulation of gene expression, and construction of internal and external developmental environments) contribute to the rise of new phenotypic variation, which may be viable and adaptive (e.g. "facilitated variation"). In addition to accepted evolutionary processes that directly change gene frequencies, i.e. processes that bias the outcome of natural selection— particularly ontogenetic bias and niche construction—are also recognized in EES. A broadened notion of inheritance encompasses genetic, epigenetic, and ecological (including cultural) inheritance. "Mutation pressure" refers to the population-level consequences of repeated mutation, which are depicted as dashed because "mutation" is also shown in "processes that generate novel variation". *1* Developmental bias and niche construction can also affect other evolutionary processes such as mutation, drift, and gene flow. *2* In EES, this category of processes will often need to be broadened to include processes that change the frequencies of other heritable resources

Numerous examples of external factors influencing early ontogenetic stages were described in West-Eberhard's [393] excellent book *Developmental Plasticity and Evolution*, and they are related to the concept of "entrenchment" that she latter designated as "environmental assimilation". According to her, "environmentally induced novelties can persist in spite of selection against them, as long as the inducing factor is present alongside individuals that respond to it; disadvantageous mutational inductions, by contrast, are likely to be eliminated by selection" [395: 114–115]. This means that "environmentally induced novelties have a better chance of persisting long enough to respond to selection and be converted into advantageous traits." Therefore, "environmental materials and cues can become entrenched in develop-ment, including in the development of constitutively expressed 'fixed' traits … ge-netic accommodation … is evolutionary change in gene frequencies due to selection on variation in the regulation, form, or side effects of a novel trait in the subpopulation of individuals that express it." It "differs from genetic assimilation in applying to both mutationally and environmentally initiated novelties, referring to both regulation and form, and applying to decreased as well as increased genotypic control." The examples compiled in her 2003 book, and the numerous other examples that Evo-Devoists have provided since then, as well as those described specifically in humans (e.g. [348]), further illustrate the difference between external selection (either at earlier stages or later developmental stages) and developmental internal selection. For more details about, and a comparison between, the Baldwin effect, Waddington's genetic assimilation, and West-Eberhard's genetic accommodation, see Chap. 8.

Internal constraints are obviously more similar within the same higher clade (thence the so-called "bodyplan"), and major behavioral shifts might not happen as often as do environment changes or habitat shifts, in at least some cases. After the first human ancestors chose to walk fully bipedally, there is no evidence that any of their descendents abandoned bipedality despite the numerous biotic and abiotic changes that our lineage experienced since then. That is, a behavioral shift can define the behavior of a last common ancestor (LCA) and thus, by way of behavioral persistence, the behavior and likely also the form—by way of, e.g. epigenetic phenomena and genetic changes that are then favorable through natural selection—of all its descendants despite the numerous environmental changes that might occur since then. Is human bipedalism really "obligate bipedalism"? Being a father of a small baby, I can attest that bipedalism is not at all "innate" or "obligate" in our species. We actively choose to teach our children to walk bipedally, and they actively choose to imitate our bipedal postures, generation after generation, and in many respects this is actually a hard work for both the parents and the babies. Would most small babies separately isolated in different islands with no other humans around "naturally" become bipeds? Surprisingly—or perhaps not because of course it would be unethical to undertake experimental studies on the subject—I could not find any detailed scientific reports on the subject. However, the available anecdotal information available does indicate that they do not, i.e. "feral" children (raised by animals) normally seem to continue to walk quadrupedally into adulthood (see, e.g. [196, 340]). Interestingly, in a clear indication of how the Neodarwinist gene-centered view of evolution (that sees organisms merely as passive actors) became so dominant within the media and general public, the discussions on such cases of feral children both online and in the press tend to argue that these are examples where the human "genetically fixed instinct" to be bipeds was lost. This is exactly the opposite of what would be the logical explanation if those anecdotal reports were confirmed scientifically. Their confirmation would indicate instead that this instinct was not lost because it was actually never acquired. What was truly acquired in humans was the behavioral persistence to walk bipedally, passed on over generations through social heredity and thus due to a continuous, active drive by members of each generation, i.e. the drive of parents to teach the babies and of the babies to learn from/imitate the parents, to walk bipedally. That is, it was that aspect of human social heredity that would be lost in those cases of feral children because obviously they could not learn from/imitate human adults in order to start walking bipedally.

In fact, in Chap. 4, I will refer to studies showing that, in at least some human indigenous tribes in which children consistently learn from/imitate their parents to climb trees from a young age, they actually become proficient in moving in trees using all four limbs as do our closest living relatives, chimpanzees, gorillas, and orangutans. Strikingly, features that are usually lost/reduced in postnatal development in our species, such as the large space between the big toe and the other toes found in late fetuses and newborns that mirrors the adult ape condition—which is related to foot prehensility—can often persist until adulthood in members of those human tribes. That is, human bipedalism does seem to a great extent to be related to our behavioral persistence. It does not seem to be merely associated to our form—which itself is mainly the result of epigenetic factors and of natural selection driven by our behavioral choice to be bipeds in the first place—and much less to each of the numerous, diverse specific ecological habitats in which we choose to live.

Recent behavioral studies are also accumulating increasing evidence showing the importance of behavioral choices and changes not only in response to direct changes in the external environment but also because of a change in the way organisms perceive their surroundings. This evidence further supports the idea that major evolutionary changes are not necessarily related to external changes/factors but may be mainly started by the organisms themselves. For instance, an experimental field study in a free-living population of song sparrows has shown that the perception of predation risk by itself resulted in behavioral shifts that were great enough to affect the entire population dynamics resulting in a 40% decrease in the number of offspring produced per year [407]. Another study has experimentally shown that the fear, or the lack thereof, that large carnivores inspire in their prey per se, independent of direct predation or even of the carnivores' physical presence, led to behavioral changes in both populations leading to a wide range of cascading effects that severely affected whole ecosystems [351].

Therefore, according to the previous text, it is now time to answer two key questions: why use the acronym ONCE specifically? And what are the main differences between ONCE and the Extended Evolutionary Synthesis that has been presented recently by some authors, particularly Evo-Devoists? Regarding the first question, although one might argue that all biological evolution is in reality organic evolution, the word "organic" is specifically meant to pay tribute to Baldwin's idea of organic selection. Because Baldwin's organic selection is only one of the many forces/factors of evolution integrated by ONCE (Fig. 1.2), I preferred to use the term "organic evolution". ONCE is intended to be an integrative, multidisciplinary view of biological evolution as a whole. Within this more encompassing view of evolution, the term "constrained" refers mainly to internal factors that constrain, help direct, and even catalyze evolutionary changes ("negative" and "positive" constraints *sensu* Gould: see Chap. 7). Such constraints were not emphasized in Baldwin's works—nor in the publications of most Neo-Darwinists for most of the last century—but they have become more emphasized in recent decades by authors such as Stephen Jay Gould and Pere Alberch and even more recently by many Evo-Devoists and developmental biologists. The use of the term "nonoptimal", and its contrast with Baldwin's and Darwin's views of evolution and with Neo-Darwinist and Evo-Devo ideas in general, was already explained previously and will be further discussed in more detail throughout this book.

Regarding the second question, I would like to emphasize a major point here. When I call attention to the importance of developmental internal factors and of behavioral choices and shifts and organic evolution in morphological macroevolution, I am not suggesting that natural (external) selection, or key Neo-Darwinist phenomena such as genetic drift and gene flow, do not play a major role in evolution. I am sure that they do, and they are all important within ONCE, particularly natural selection as explained previously and as shown in Fig. 1.2 (see also Chaps. 8 and 9). However, a major differences between ONCE (Fig. 1.2) and the Extended Evolutionary Synthesis —as proposed by authors such as Pigliucci and Müller [293] and Laland et al. [218] and schematized in Fig. 1.3—is that in the latter synthesis there is no special focus on long-term macroevolutionary trends and even less focus on ecomorphological mis-matches. In fact, apart from the recognition of strong developmental constraints and some randomness in, e.g. drift and possibly in gene flow, in a way that synthesis would predict a general tendency for organisms to display eco-morphological matches. This is because it proposes a direct link between environmental changes and developmental processes and gene expression, and so on (Fig. 1.3). These latter aspects are also part of ONCE, but ONCE focuses more on the importance of behavioral persistence as a main driver of evolution and also attributes a more crucial role to constraints. For instance, ONCE highlights the power of the "logic of monsters" of Pere Alberch and the links between variants, defects, and normal evolution as well as emphasizes the potentially important role that Goldschmidt's hopeful monsters and saltational evo-lution might have in macroevolution (Fig. 1.2; Chap. 8).

Another major difference between ONCE and the Extended Evolutionary Synthesis is the crucial role played by randomness in ONCE (Fig. 1.2). This role includes phenomena such as drift/gene flow in particular, as does the latter syn-thesis, but in addition it takes into account the particularly crucial role that be-havioral choices play in ONCE. Even within the same external environment and in response to the same factors, organisms can make an impressive number of different behavioral choices. Therefore, they might not always make the most "logical" choice for the habitat in which they live, potentially resulting in etho-ecological mismatches, as explained previously. For example, if a group of birds migrates to a region that has dozens or hundreds of islands, the first island that they see will be a random one. It is not necessarily the "best" one, and they might decide not visit or stay on it. However, simply by being the first island the birds saw, it would actually play a huge role in that behavioral choice and thus in the subsequent choices and the future of the bird population, by mere contingency. If they decide to visit/stay on the island, conditions might prove to be difficult over time, and the birds might migrate again; however, in many cases they might be able to stay on the first island, particularly through niche construction and the help of phenomena such as epige-netic factors directly related to the conditions/factors present in that island, and so on. In this way, ONCE emphasizes the importance of contingency and of ran-domness—incorporating Gould's thoughts on the subject as will be explained just below—to an even higher level than does the Extended Evolutionary Synthesis, by focusing more on the frequent occurrence of both ecomorphological mismatches and macroevolutionary trends (Fig. 1.2) than does this latter synthesis (Fig. 1.3)

The influence of past history and randomness on evolution was emphasized in Gould's metaphor about "replaying life's tape": "you press the rewind button and, making sure you thoroughly erase everything that actually happened, go back to any time and place in the past; then let the tape run again and see if the repetition looks at all like the original; any replay of the tape would lead evolution down a pathway radically different from the road actually taken" [157]. This subject has been widely discussed, theoretically and empirically, both before and after Gould's publications. A recent article by Plucain et al. [295], and the references therein, provide some good examples of such studies and current discussions on this topic. They noted that two main approaches are often used to investigate historical contingency. The first approach analyzes the level of evolutionary parallelism at the macroevolutionary level, e.g. whether major novelties that have repeatedly evolved in evolution or whether evolution are repeatable in natural populations/species that evolved independently in similar but geographically disconnected environmental conditions such as oceanic islands or glacial lakes. An empirical example they review involves the repeated adaptive radiations of *Anolis* lizards in Caribbean islands.

According to Plucain et al. [295], the second approach investigates historical contingency by experimental evolution with viruses, bacteria, and yeasts that are propagated from an initial ancestral strain under laboratory conditions for hundreds to thousands of generations. They note that several studies showed that contingency might play a major role in the evolution of protein structure and properties and reported rare phenotypic novelties that required consecutive mutations and a potentiating genetic background. For instance, the emergence, after a very long evolutionary time ($> 30,000$ generations), of a citrate-utilizing phenotype in *Escherichia coli* required a sequence of mutations accumulated during a specific moment of their evolution. Likewise, a particular bacterial polymorphism was contingent on the complex interactions between different mutations. However, they note that other studies pointed out that adaptation is not always historically contingent. For example, researchers compared fitness-related phenotypes in strains of *E. coli* with different initial genetic conditions and evolutionary histories and found them to be similar. They further noted that phenotypic adaptation of Tobacco virus strains to their *Nicotiana tabacum* host also proved to be independent of past evolutionary histories. Similarly, yeast populations have been shown to converge at the phenotypic level and partially at the genomic level during adaptations.

However, as Moran [264: 1] pointed out, some confusion exists between contingency and chance/randomness in the literature. As he notes, "contingency is not the same as chance or randomness; contingency refers to historical events where each step is dependent upon, or contingent upon, preceding events; the final result of any long historical process is the product of a large number of distinct circumstances any one of which might have been different." If "all these circumstances were severely constrained, then the end result might be highly predictable, like the outcome of a computer algorithm; on the other hand, if most of the circumstances were the result of random events, like lucky accidents, then the end result would be unpredictable." Therefore, when authors such as Moran [264] refer to evolution by accident—often in metaphysical discussions about the existence of God and other similar topics, which are not the main focus of the present book—they are "referring, in part, to a history of evolution that includes a huge number of chance and random events all of which are contingent upon everything that preceded them."

Bonner's [45] book, *Randomness in Evolution*, provides an updated, short, and elegant discussion of randomness including some interesting thoughts on the links between randomness, size, and time. He considers that in early evolution small and relatively simple organisms were mainly affected by randomness with natural selection playing a less prominent role. In contrast, in modern organisms that are, on average, of larger size and greater complexity, a more prominent role is played by internal selection (part of internal constraints sensu the present book: see above and Chap. 7) during development and of natural selection in mature animals and plants. I am not sure that I completely agree with this view because if we were to follow a systems biology approach (e.g. [50]), one would see the evolution of "complexity" leading to a scenario in which the homeostasis/autonomy of organisms tends to increase and thus depend less on both randomness (as predicted by Bonner) and external factors related to, e.g. changes in the external environment. For instance, humans are large, and in some ways/anatomical regions we are fairly complex. However, I am not sure one can say—as Bonner suggested—that we are more affected by external (natural) selection than animals such as frogs, with our medicine, buildings, tools, and so on. In the overall picture, according to ONCE the effect of randomness on increasingly complex organisms could be either smaller or greater. On the one hand, more complexity and more homeostasis/constraints can decrease the effect of randomness as defended by Bonner. On the other hand, if we consider that more complexity would lead, in general, to a wider range of possible behavioral choices for the organisms, then I would argue, under ONCE, that this would lead to more randomness due to the exponential growth of the overall number of possible behavioral options and their combinations.

In terms of anatomy, one interesting case study—although mainly anecdotal, because of the lack of similar case studies to which it can be compared—comes from primates. In 'phylogenetically and morphologically basal' primates—for details on the use of these terms, and on how they are *not* teleological, see Diogo et al. [109]—such as bush babies (*Galago*) one can find only two main types of disposition (variation) of the mylohyoid fissure (a fissure of the mandible, related to the mylohyoid muscle that inserts onto this bone): one occurs in about 25% of the individuals, and the other in about 75% of the individuals [411, p. 39]. In contrast, in 'phylogenetically and morphologically derived' primates such as apes there are four main types of disposition of this fissure: for instance, in chimpanzees types 1, 2, 3 and 4 are found in about 9, 32, 51 and 7% of the cases, respectively. We can thus interpret this case study as indicating that there was a phylogenetic increase of homeostasis from early primates to apes, in the sense that apes have more intra-specific variations but are able to cope with them and still have a functional musculoskeletal system. This idea goes in line with the view defended by Gilbert and Epel [151], who stated that robustness is often the outcome of an increase of complexity, because the more plasticity, the more organisms are protected against external changes (homeostasis), thus connecting robustness, homeostasis, plasticity, variation, and complexity. Be that as it may, I certainly agree with the main, and more crucial, idea of Bonner's book: randomness plays a major role in evolution. I thus refer readers to his excellent book, and the references therein, for more details on this topic.

Overall, the major difference between ONCE (Fig. 1.2) and the Extended Evolutionary Synthesis, as well as most other evolutionary theories defended by other authors—with few exceptions, such as Baldwin's organic selection—is however that in ONCE organisms themselves, as a whole, and in particular their behavioral choices and persistence, are the key drivers of evolution. Although Laland et al. [218] stated that the Extended Evolutionary Synthesis is organism-centered, in their scheme (Fig. 1.3) the behavior of organisms themselves is not even shown, being only possibly involved with niche construction, which itself does not appear as a central phenomenon within that scheme. As will be further discussed in the chapters below, it is in fact surprising that in that synthesis, and even in some specific works within the niche construction literature, organisms often continue to be mainly seen as passive players. This tendency is well illustrated by a sentence from the main discussion of Schlichting and Pigliucci's [327: 183, 312–323] book: "perhaps the most general message to emerge from this potpourri of examples is that the discourse among genes, characters, and the environment can easily lead to complex patterns of phenotypic expression." Even if one argues that the "characters" include behavioral traits—which is almost never done explicitly in that book—this still represents a very non-organismal, non-behavioral view of evolution. This non-organismal view is particularly striking because they do recognize that a review of the empirical data accumulated in the last decades of the last century—and since then, I would add—seems "to elevate genetic assimilation to a position where it is the most likely mode of evolutionary change for many characters." Moreover, the authors recognize that "the concept of genetic assimilation has an extended history, dating at least to the writings of Baldwin [24–26], Osborn [286], and Lloyd Morgan [231] on what they referred to as 'organic selection'... its present incarnation emerged from the work of Waddington, who coined the phrase, and Schmalhausen, who referred to it as 'stabilizing selection'."

However, despite all of these correct associations, Schlichting and Pigliucci do not focus on the key aspect linking them, i.e. organic selection, and thus the behavior (habits) of the organisms themselves. The term "behavior" is not even among the hundreds of items listed in the index of their book, which, as its title, *Phenotypic Evolution—A Reaction Norm Perspective* indicates, deals with topics that should be profoundly related to behavioral evolution. Thus, my question is this: where are the organisms themselves, their behavioral choices, the systems biology notion of emergence—that the sum is more than the parts—within that discourse about "genes, characters, and the environment"? Do organisms—as a whole, as full living beings—play any role in that discourse? Or are they merely passive carriers of "characters" and "genes," going with the flow created exclusively by the interplay between internal characteristics and constraints and external environmental forces? Within the context of ONCE, the opening sentence of the present book, a quote from Thomas Muggeridge that was originally embedded in religion—when The Right Reverend Robert Daly stated, in 1826, "live fish swim against the stream, while dead ones float with it"—is key: "never forget that only dead fish swim with the stream". That is, the answer is no: living organisms do not simply go with the flow.

Chapter 2
Baldwin's Organic Selection and the Increasing Awareness of the Evolutionary Importance of Behavioral Shifts

Baldwin's developmental and evolutionary theories of organic selection were developed in the context of passionate debates about a wide range of broader biological and evolutionary topics, approximately 120 years ago (e.g., [23–26]). Baldwin's main aim was to explain how directional evolution could occur without the Lamarckian direct inheritance of acquired characters. It is therefore very interesting to see how the so-called "Baldwin effect" has been increasingly widely discussed in the last decades. In addition, in recent years evolutionary biologists from very different backgrounds—including psychology, ethology, ecology, and developmental biology—have also become increasingly interested in Baldwin's broader idea of "organic selection". A quick search in Google Scholar shows that, since 2000, at least 735 publications have focused on "organic selection" and up to 4360 works have focused on the "Baldwin effect". Due to a lack of space, and to not lose the focus on its central message, I will not provide an extensive account of Baldwin's life and work in this chapter. This is beautifully done in many of those thousands of publications, including entire books dedicated to this and related subjects such as *Evolution and Learning—The Baldwin Effect Reconsidered* [390] and *Beyond Mechanism: Putting Life Back Into Biology* [182]. I will simply provide a brief summary of those of Baldwin's key ideas that are particularly relevant to ONCE. Part of this summary will be based on a recent paper by a fascinating and outside-the-box thinker who provides a well-considered introduction to Baldwin's organic selection, its historical context and influences, and the differences between his ideas and those of Lamarck and Darwin: Corning [61].

Corning notes that both Lamarck and Darwin appreciated that functional adaptation to an environment is problematic for organisms. However, Lamarck argued that because the environment is not fixed, when it changes organisms must accommodate themselves or they will not survive/reproduce. Changes in the environment over time can thus lead to new "needs", which in turn can stimulate the adoption of new "habits": thus, in theory, changes in habits (function/behavior) would come before changes in structure (form). As noted by Corning, Darwin also valued the role of behavior in evolutionary change but was more cautious, stating

R. Diogo, *Evolution Driven by Organismal Behavior*,
DOI 10.1007/978-3-319-47581-3_2

that "it is difficult to tell, and immaterial for us, whether habits generally change first and structure afterwards; or whether slight modifications of structure lead to changed habits; both probably often change almost simultaneously" [67: 215]. Moreover, most of Darwin's followers did not acknowledge an important evolutionary role for behavior, except for some scientists who "Darwinized Lamarckism" at the turn of the twentieth century. These include, among others, Baldwin, Morgan, and Osborn: despite their different individual perspectives, their overall views were generally lumped together under Baldwin's Organic Selection [61].

Corning wisely uses the most common example to differentiate the theories of Lamarck and Darwin, i.e. the size of giraffes and their necks, from Baldwin's organic selection. According to Baldwin's view, naturally occurring variations in the neck lengths of ancestral Giraffidae (i.e., the family that includes extant giraffes) would have become adaptively significant when these animals acquired, probably through behavioral trial-and-error, a new "habit" (eating Acacia leaves) as a way of surviving in the relatively dry environment of the African savanna. As stressed by Corning, we cannot know for sure that this was the case, but some suggestive evidence can be found in another species of the family Giraffidae, the okapi, which occupies woodland environments. As expected within the context of Baldwin's theory, the members of this species have very different feeding habits than giraffes do and accordingly also have a different form: their necks are much shorter than those of giraffes (Fig. 2.1). Paleontological studies in fact suggest that, compared with their last common ancestor, neck length was both dramatically increased in the giraffe lineage and secondarily decreased in okapi lineage [66]. This example therefore illustrates how behavioral choices—such as whether to move elsewhere or continue to live in a savannah environment versus a woodland one, and specifically whether to eat Acacia leaves in the former environment or leaves, buds, and shoots in the latter—can lead to very different evolutionary trends in giraffes compared with okapis (i.e. increase vs. decrease of neck size).

This example also shows how Baldwin's ideas, *Neo-Darwinism*, and Lamarckism share some common points but also have fundamental differences. Neo-Darwinism would mainly argue that the first, and crucial, changes were random mutations that resulted in giraffes having both longer and shorter necks and thus allowed those with longer necks to eat Acacia leaves. That is, that the behavior of the giraffes themselves, per se, was mainly a secondary actor in the story. Baldwin's idea is similar in the sense that mutations are needed to explain the directional evolution leading to a longer neck, but the first key driver was the behavioral choice of the giraffe ancestors to move to, or continue to live in, a dry environment and then to eat certain specific types of plants and ultimately Acacia leaves, among the numerous other plant species existing in the African savanna. Only *subsequently*, within this specific behavioral context and constructed niche, did certain random mutations specifically providing advantages for this mode of life thrive. This view suggests a codependent process between the active behavioral choices of organisms, genetic phenomena such as random mutations, and natural (external) selection.

According to Lamarckism, the first key actors of the story are also the giraffe ancestors and their behavioral choices, as for Baldwin. Lamarck's ideas were also

Fig. 2.1 Skeletons of an okapi (*left*) and a giraffe (*right*), which are both from the family Giraffidae (modified from [74])

similar to those of Baldwin—although, as explained previously, the latter did not recognize this point—in the assumption that acquired traits are inherited, in the sense that the behavioral persistence to eat Acacia leaves is passed down through the generations. That is, it is inherited through teaching, learning, imitation, and so on (social heredity *sensu* Baldwin) and possibly then partially 'fixed' in giraffes by way of the Baldwin effect. The main difference is the way in which Lamarck used, his concept of "use-and-disuse" as though a giraffe could, over a lifetime of effort to reach high branches, develop an elongated neck and pass that acquired morphological feature to its descendants. Such an idea is of course essentially wrong, and it is the main factor that has led many authors to even avoid using Lamarck's name. However, it is clear that Lamarck's thoughts were much more complex, and in general more accurate, than this caricature. His key idea of inheritance of acquired traits does apply to behavioral and ecological inheritance, for instance, as is now widely recognized (Figs. 1.2 and 1.3) and as will be discussed in more detail in Chap. 5.

Box—History: Baldwin and the debate on preformationism versus epigenesis
As stressed by Kull [215] and Young (2003), Baldwin essentially opposed a strict preformationist position. In a very simplified way, preformationists argued that organisms are preformed, i.e. that they develop from miniature versions of themselves, and therefore that their ontogeny is mainly defined from the beginning by a specific, defined program. Young [413] summarized

a few contrasting tenets of epigenesis versus preformationism. In the former, variations appear in definite directions because they are caused by the interaction of the organic being and its environment, and thus they can be inherited. In the latter, variations are promiscuous, being "congenital" or caused by mixing of male and female germ-plasmas, and thus could not be inherited. Contrary to the suggestion of many authors that attacked him during the first decades of the twentieth century, Baldwin was strongly opposed to a Lamarckian epigenesist position. For Baldwin, the Lamarckian inheritance of acquired characters was too rigid because it would not allow organisms to display behavioral plasticity and accommodate themselves to the external environment and/or conditions in order to alter their behavior in adaptive ways (Young 2003). Baldwin (e.g., [23–26]) argued that plasticity (both behavioral and morphological) is essential for the new behavior to be acquired and performed and for its later imitation and/or learning by other members of the population. This in turn allows social heredity, i.e. the passage of behaviors from one generation to another, which works in concert with external natural selection so that the more useful behaviors/adapted phenotypes are selected across generations.

Baldwin [25: 549–552] stated that "natural selection is too often treated as a positive agency; it is not a positive agency; it is entirely negative… it is simply a statement of what occurs when an organism does not have the qualifications necessary to enable it to survive in given conditions of life; it does not in any way define positively the qualifications which do enable other organisms to survive. He added:" … assuming the principle of natural selection in any case, and saying that, according to it, if an organism does not have the necessary qualifications it will be killed off, it still remains in that instance to find what the qualifications are which this organism is to have if it is to be kept alive… so we may say that the means of survival is always an additional question to the negative statement of the operation of natural selection." He explained that the term "organic" in organic selection is related to "the fact that the organism itself cooperates in the formation of the adaptations which are effected, and also from the fact that, in the results, the organism is itself selected; since those organisms which do not secure the adaptations fall by the principle of natural selection" [25]. For him, the word "selection" is "appropriate for just the same two reasons… animals may be kept alive let us say in a given environment by social cooperation only; these transmit this social type of variation to posterity; thus social adaptation sets the direction of physical phylogeny and physical heredity is determined in part by this factor."

Significantly, the term "social heredity", as used by Baldwin and in the present work, does not apply only to taxa in which individuals display a highly complex social organization and/or neurobiological skills. "Social heredity" is simply one of the factors that can lead to behavioral persistence—or behavioral inheritance—for Evo-Devoists (Fig. 1.3) through e.g. teaching, learning, or imitation, but this can be done at a very basic level without necessarily invoking consciousness or any particularly complex type of behavior. As Birkhead and Monaghan [38: 12] stated, in behavioral ecology "animals are viewed as having *choices*", and the use of such terms caused misunderstandings and "provided ammunition for those opposed to behavioral ecology", for whom such terms "implied a conscious decision by an animal - something the behavioral ecologists never intended." Or, as stated by Ydenberg [412: 132], "in behavioral ecology the term *decision* is used whenever one or two (or more) options is/are selected, with no implication that the choice is conscious; the choice need not be cognitive at all, and may not even use neural mechanisms." Basically, any organism can thus make behavioral choices, as I argue in the present book. As will be explained later in the text, in this sense Baldwin's "social heredity" has similarities to factors leading to the so-called "ecological inheritance" of the niche construction theory and thus can be applied to any type of biological organisms from bacteria to modern humans, elephants, dolphins, or octopuses. Another important point is that "plasticity" here refers to different possible alternatives, e.g. as noted previously behavioral choices/shifts will not always "meet the needs" of the organism; they can also go wrong and be subsequently eliminated by natural selection.

As summarized by Kull [215], according to Baldwin's idea, a population might thus first face new conditions (e.g. due to change in their environment either locally or due to migration), and then its organisms *accommodate* to the new conditions by way of physiological adjustment/behavioral changes, possibly due to physiological/behavioral plasticity. Significantly, when a population faces a change in conditions, all of its organisms may respond simultaneously and in a similar way. Second, the new behavior and/or niche built by the organisms can last for generations because of the permanence of the new habitat conditions, the stability of the environmental conditions, the continuation of a newly established ecological link with other organisms or food resources, and/or, very importantly for Baldwin, because of behavioral persistence due to learning/imitation or other factors. Third, natural (external) selection then plays a crucial role because it allows a trial-and-error type of evolution, in which random mutations that lead to behavioral/morphological adaptations within the context of the new behavior/habitat can be selected (while others are not), thus explaining the occurrence of directional evolution/evolutionary trends that last for long geological time periods. Therefore, while abandoning the transmission of acquired characters, Baldwin's organic selection did place individual behavior and adaptation first and random mutations/variations second, as Lamarckians contended, instead of placing survival conditions by fortuitous mutations/variations first and foremost as NeoDarwinists defend.

According to Hoffmeyer and Kull [188: 263–265], a major difference between Neo-Darwinism and Baldwin's organic selection is that in the former "an ability to use sign processes", e.g. in behavioral choices/shifts, may "turn out to be an advantage in the struggle for existence (like many other features, such as an ability to move quickly), but it cannot itself be a factor that is sufficient for *creating* evolutionary adaptations" as proposed by Baldwin. Hoffmeyer and Kull reviewed an illustrative example of a behavioral shift of a population of invertebrates in which there was no genomic change: "the dreaded locust, which most of the time lives its life as an ordinary, harmless grasshopper", but which, under certain conditions, changes its behavior. This change leads to "new generations with a markedly changed morphology and behavior, causing these locusts to form enormous flocks flying many kilometers, and devouring every green thing in their path." For them, such examples stress how behavioral shifts can take place simultaneously in many individuals of a population (as a result of a change in environment or migration to a new environment) in contrast to the difficulty of explaining how random mutations can spread so quickly throughout an entire population. The behavioral shift, and the subsequent phenotypic shifts to which it potentially leads, may be sufficient to decrease the effectiveness of recognition of the original population needed for mating, thus leading to isolation and possibly to subsequent mutations that will fix this separation also at the level of the genome/other type of incompatibility. This idea therefore links behavioral shifts, speciation, and cladogenesis. In fact, as stressed by Larsen [220: 120], the "migratory locust form differs from the solitary form in a variety of ways; not only is behavior modified, but pigmentation and morphology as well; changes in phase do not occur in one generation but require several reinforcing generations in which maternal effects are important, since it appears that maternal juvenile hormone influences juvenile hormone titers of their final instar progeny."

Many of the previous points were elegantly summarized in Kull's [215] table, which I use and update here (Table 2.1) to summarize the main differences between ONCE and the views of Baldwin, Neo-Darwinists, Lamarckians, and mutationists such as Morgan. As shown in Table 2.1, a crucial point of Baldwin's organic selection is that the first event is a plastic/phenotypic change, which is followed by stochastic genetic changes. This view contrasts with (1) Lamarckism (i.e. first event is also a plastic/phenotypic change but is followed by the inheritance of acquired characters), (2) mutationism (i.e. nonrandom mutations with genetic change

Table 2.1 Different views on adaptive evolution

	Nonrandom mutations/epigenetic phenotypical events	Random mutations	Random mutations but also epigenetic phenotypical events selected/influenced by external selection, both being strongly constrained by internal factors
Epigenetic changes (e.g. learning) first	Lamarckian (e.g. exercising/use-disuse)	Baldwinian/semiotic (organic selection)	Organic nonoptimal constrained evolution (ONCE [the view defended here])
Genetic changes (e.g. mutations) first	T. H. Morganian (mutationism)	Neo-Darwinian (natural selection)	–

occurring first), (3) Neo-Darwinism (i.e. the first event is a random genetic change followed by a new phenotype and natural selection), and (4) nonadaptive, neutral, and/or other mechanisms of evolution such as genetic drift. As will be explained later in the text, it also contrasts with ONCE because Baldwin's organic selection is just one of the major points of ONCE. For instance, ONCE also incorporates data from recent Evo-Devo studies stressing the influence of the external environment on multiple aspects of early development through epigenetic factors that were not known in Baldwin's time as well as studies on the links between ecology, morphology, and phylogeny showing that etho-ecological and eco-morphological mismatches are far more frequent than Baldwin's idea would predict (Fig. 1.2). I will also explain, in further chapters, that Baldwin's criticism of Lamarck's inheritance of acquired traits is in fact invalid if one considers that these traits do include behavioral/ecological features. Of course, one can argue that behavioral persistence, in the way Baldwin defined it, might not be the same as the "inheritance of behavioral traits" because those traits are not innate: within the general context of organic selection they must be gained over and over again, during each generation, e.g. through teaching, learning, or imitation. However, within this general context, even Baldwin defined a subset of cases that can in fact lead to innate behavior, which became known as the "Baldwin effect" (see Chap. 1 and later text).

Due to the emergence of genetics, Baldwin's ideas—as well as Lamarckism— became largely ignored, but this downturn was transient. For instance, Simpson [338] renamed, and disseminated, this "Baldwin effect" component of Baldwin's main idea of organic selection [61]. In contrast, Waddington, who published several works in the 1940s and 1950s, strongly criticized the gene-centered Neo-Darwinist view of evolution. He stated: "it is the animal's behavior which to a considerable extent determines the nature of the environment to which it will submit itself and the character of the selective forces with which it will consent to wrestle; this 'feedback' or circularity in a relation between an animal and its environment is rather generally neglected in present-day evolutionary theorizing" [384: 170]. A very interesting point made by Corning [61]—which, in my opinion, is not emphasized enough in the literature—is that a major reassessment of Baldwin's organic selection did occur in the late 1950s. Specifically, this occurred when the American Psychological Association and the Society for the Study of Evolution jointly organized various conferences that resulted in the book *Behavior and Evolution,* edited by Roe and Simpson [311]. The book included the following suggestions: (1) adaptive radiations might be fundamentally behavioral in nature (Simpson), (2) behavior might often serve as an isolating mechanism in the formation of new species (Spieth), and (3) during evolutionary transitions, new behaviors may appear first and genetic changes may follow (Mayr). As noted by Corning, Mayr later wrote, in his influential chapter "The Emergence of Evolutionary Novelties", that behavioral changes are the "pacemaker" of evolution [247], an idea that he also discussed in his 1976 book *Evolution and the Diversity of Life.*

In the late 1950s and 1960s, there was a dramatic increase in the number of publications highlighting the role of learning and behavior in evolution, including Thorpe's [353] *Learning and Instinct in Animals,* Waddington's [380, 383] *The*

Strategy of the Genes and The Nature of Life, Hardy's [174] *The Living Stream*, Whyte's [398] *Internal Factors in Evolution*, Hinde's [184] *Animal Behavior: A Synthesis of Ethology and Comparative Psychology*, and Koestler's [210] *The Ghost in the Machine* (see [61]). Since then, research on learning and innovation in living organisms—from "smart bacteria" to wise apes and playful dolphins—has grown exponentially. This includes empirical data suggesting that *Escherichia coli* bacteria, *Drosophila* flies, ants, bees, flatworms, laboratory mice, pigeons, guppies, cuttlefish, octopuses, dolphins, gorillas, and chimpanzees—among many other taxa—can learn and display novel responses to new conditions by way of, e.g., "classical" and "operant" conditioning. According to Duckworth [117: 414], encompassed within this growing recognition that behavior is crucial in evolution, are two contrasting ideas about how behavioral changes affect evolutionary rates.

On the one hand, behavior can be seen as a constraint in the sense that it could in theory slow the rate of evolutionary change because behavioral plasticity can shield organisms from strong directional selection by allowing them to either exploit new resources or move to a less stressful environment. For instance, Morris [265: 2–3] noted that "plasticity, by slowing the rate of population decline, can overcome this hurdle, providing time for beneficial mutations to arise; this has been confirmed empirically: the likelihood of extinction for great tits increased 500-fold in the absence of plastic responses to climate change… this was largely true in thirteen other bird species, although faster generation times offset the need for plasticity." Also, "climate change has resulted in population declines of numerous nonplastic species… for instance, rising temperatures shifted flowering time but not West Greenland caribou calving time, producing a trophic mismatch that declined calf production fourfold." Morris further noted that "in order to successfully invade a new environment, individuals must first disperse to that environment… species with high dispersal rates should also be highly plastic, as dispersal involves encountering spatial heterogeneity… indeed, dispersal of nonplastic organisms can reduce the likelihood of successful colonization by introducing maladaptive alleles to colonizing populations." He reviews a study that tested the relationship between dispersal and plasticity in 258 species of marine invertebrates and showed that, on average, dispersing species were more plastic than nondispersing ones, i.e., presumably without such plasticity dispersers would fail to colonize the locations to which they disperse. However, he noted, "it is clear that many species have high dispersal rates and low levels of plasticity, so plasticity is again sufficient but not necessary for colonization; rapid generation times, for instance, may allow colonizing populations to rapidly evolve to meet the demands of the new environment." On the other hand, it can be said that even in such cases behavior would actually drive evolutionary change. A behavioral shift that results in a new way of interacting with the existing environment, or a switch to a new environment, exposes organisms to novel selection pressures resulting in evolution of life history, physiology, and morphology as emphasized by Baldwin and in ONCE (Fig. 1.2). In fact, as stressed by Duckworth [117: 414], the major point is that, be that as it may, these examples "emphasize that the critical novel step in the evolutionary sequence is a behavioral shift."

Interestingly, in recent years Baldwin's organic selection has become particularly fashionable in the fields of systems biology and complex systems because it is closely related to a term currently very much in vogue: "teleonomy". Within these fields, this term, coined by Pittendrigh in his chapter "Behavior and Evolution" of Roe and Simpson's [311] book, refers to the "internal teleology", i.e. "the fact that the purposefulness found in nature is a product of evolution and not of a grand design" [61: 248]. As Corning explains, there are historical links between this term and not only 'Intelligent Selection' *sensu* Morgan and 'Organic Selection' *sensu* Baldwin but also 'Holistic Selection' *sensu* Smuts and 'Internal Selection' *sensu* Whyte and Koestler in the 1960s and, more recently, with 'Psychological Selection' *sensu* Mundinger, 'Rational Pre-Selection' and 'Purposive Selection' *sensu* Boehm, 'Baldwinian Selection' *sensu* Deacon, 'Neo-Lamarckian Evolution' *sensu* Jablonka and Lamb, 'Behavioral Selection' *sensu* various authors, 'Selection by Consequences' *sensu* Skinner, and 'Social Selection' *sensu* biological anthropologists. The crucial common link between these concepts is that living beings do the selecting: they have emergent properties that allow problem-solving, innovation, and decision-making so they can—and do—choose among various possible behavioral options.

For Corning, the proximate causes of novel forms of symbiosis—from lichens to such evolutionary turning points as the origin of eukaryotic cells as well as the origin of land plants and animals, the evolution of birds, and even the rise of social organization—were most likely the result of various behavioral 'initiatives.' One emblematic example of organic or "teleonomic" selection he provided is the intense competition among the towering evergreen trees (western hemlock, Sitka spruce, Douglas fir, and western cedar) in a forest canopy of the rainforest of the Olympic National Park in Washington State. Hemlocks produce by far the most seeds and are said to be the best adapted to grow in the park as an outcome of both competition and the weather, especially the low-sunlight conditions. However, the Sitka spruce dominates because the abundant Roosevelt elks in the park feed intensively on young hemlock trees but do not eat Sitka spruces. That is, the food preference of the elk is the *proximate cause* of differential survival between the hemlock and spruce trees. Similarly, the many kinds of artificial selection practiced by humans, including sexual selection, can be seen as behavioral selections by third parties that shape the course of natural selection in other species as noted in Chap. 1. I will discuss sexual selection in some detail in Chap. 4.

Chapter 3
Behavioral Choices and Shifts, Niche Construction, Natural Selection, Extinctions, and Asymmetry

As noted in Chap. 1, in the last few decades, particularly with the rise of Evo-Devo, authors have increasingly pointed out the need for a "post-Darwinian view" of evolution or an Extended Evolutionary Synthesis (Fig. 1.3) (e.g. [215, 293]). These authors usually dispute the gene-centered view of evolution of Neo-Darwinists. For instance, some authors attempt to make the "epigenetic turn", stressing the importance of cellular, physiological or anatomical traits that are mainly related to external/environmental factors, and not exclusively "coded" by the genome (e.g., [198]). Such attempts revived some ideas that were historically associated—rightly or wrongly, but usually with a negative connotation—with Lamarckism, and therefore with the epigeneticist position within the preformationism *versus* epigenesis debate, which was a particularly hot topic in the nineteenth century [413]. As explained in Chap. 2, in a way the idea of preformation has some similarities with the Neo-Darwinian, gene-centered view of evolution, in which the phenotype is basically said to be coded in the genotype.

Authors such as West-Eberhard [393] have been particularly influential in promoting the importance of epigenetics in evolution by arguing that the external environment profoundly affects even early developmental stages, e.g. her concept of "entrenchment" (see Chap. 1 and, e.g., [84, 234]). Other authors are also attacking the gene-centered view of Neo-Darwinists but in a different way: they focus mainly on internal developmental factors, such as the physical forces that are crucial within the physicalist framework *sensu* authors such as Newman and Müller (see also Chap. 1). In recent years ecological inheritance has been widely recognized as a core component of extra-genetic inheritance and has become central to attempts within evolutionary biology to broaden the concept of heredity beyond transmission genetics (Fig. 1.3; [219]). According to this idea, the development of organisms, as well as the recurrence of traits across generations, crucially depends on the construction of developmental environments by ancestral organisms.

In this sense, niche construction is not simply a source of environmental change but rather a driver of selection that may produce novel evolutionary outcomes. For instance, when an organism builds a nest it creates selection for the nest to be

© Springer International Publishing AG 2017
R. Diogo, *Evolution Driven by Organismal Behavior*,
DOI 10.1007/978-3-319-47581-3_3

defended, maintained, and regulated, as well as for other organisms to inhabit or dump eggs in it. Therefore, as stressed by Laland et al. [219], it is important not only to present examples of niche construction, such as the nests and burrows of animals, but also to stress that in such cases the constructing activity has fed back to influence the evolution of the "constructor" or any other taxa. Laland et al. provide examples of recent works that clearly demonstrate this evolutionary feedback, including studies that have experimentally shown that niche construction can evolve rapidly, under a broad range of conditions in microbial populations. For example, yeast can modify its fruit environment to attract the fly *Drosophila*, thus facilitating its own propagation, and there is also a feedback between the ecosystem-modifying activities of adult fish and the fitness relationships among juveniles in a subsequent generation.

An example of the links between niche construction, contingency, ecological associations, radiations, and macroevolution that I consider particularly useful is reviewed by Erwin [124]. Lucinid bivalves originated in the Silurian (a period extending from approximately 444 to 419 million years ago). However, their diversity remained low until the Cretaceous period (145–66 million years ago) when they began a notable radiation that persisted until the Cenozoic period (66 million years ago to the present). This Cenozoic diversification was mainly related to the "construction", by seagrasses and mangroves of a habitat of dysaerobic (in which oxygen is less than normal [aerobic] although not zero [anaerobic]) sediments below their root zones. Lucinid bivalves contain endosymbiotic bacteria in their gills, which thrive on the sulfide generated in the dysaerobic sediments. In turn, the seagrasses benefited from the reduction in sulfides. The bivalves possessed the features that were, in theory, required for this diversification when they arose in the Silurian, but the habitat that became so beneficial for their diversification did not exist at that time. Therefore, as noted by Erwin, this is an example of evolutionary diversification linked to niche construction, in which modification of a niche by the members of one taxon alters the ecological, and in this case evolutionary, opportunities for the organisms of another clade.

The recent revival of Baldwin's idea of organic selection has been in great part due to an increase in the number of authors defending the notion and importance of niche construction. As put by Griffiths [161: 204–209], "niche construction fundamentally influences evolutionary dynamics because it implies that organisms are not so much adapted to their niches as coevolved with them." That is, according to authors such as Lewontin, "organisms both physically shape their environments and determine which factors in the external environment are relevant to their evolution, thus assembling some subset of the biotic and abiotic factors in their physical environment into a niche; organisms are adapted to their ways of life because organisms and their way of life were constructed for (and by) one another" (see Fig. 2.1). Laland et al. [217] recently argued that the evolutionary significance of niche construction stems from the fact that (1) organisms modify environmental states in non-random ways, thus imposing systematic biases on the selection pressures they generate; (2) because organisms change the environments of their descendants, niche construction creates a supplementary kind of inheritance

("ecological inheritance"); (3) "acquired" characters become evolutionarily significant by changing their selective environments; and (4) the match between
organisms and their environments (historically designated as "adaptation") can thus
be increased by way of niche construction (changing habitats to suit organisms), not
merely by way of natural (external) selection.

Laland et al. provided the following example, which further emphasizes the
importance of behavioral shifts in evolutionary change, to back their argument.
Earthworms evolved from freshwater oligochaetes, but they are now terrestrial, not
so much because of major changes in their form but instead because they restructure
the soil environment to increase water retention and accessibility. For instance, by
digging and tunneling they increase the incidence of soil macropores, thus promo-
tion the infiltration and capture of rain water. For example, their activities reduce the
soil's clay fraction, thus weakening soil matrix potentials and making soil water
more accessible to the worms, and also moderate the soil's daily march of temperature. As noted by the authors, such a restructuring of the soil environment
creates an interface more suited for the earthworms' essentially aquatic physiology.
In turn, this restructuring alters the selective regime for earthworms and their
descendants, as well as that of many other organisms that inhabit the same soils. That
is, because the modifications on the soil performed by earthworms last longer than an
earthworm's typical life span, this constructed niche can lead to a form of "external
hereditary memory", i.e., to ecological inheritance *sensu* ONCE (Fig. 1.2).

Another illustrative example of niche construction that I particularly like was
reviewed by Turner [362], who moreover provided a niche scheme to summarize
the main differences between a traditional, more gene-centered view of evolution
and the more recent notion of niche construction (Fig. 3.1). The singing burrows of
mole crickets (*Gryllotalpa* spp.) are produced into an exponential horn that can
broadcast a mating call. The burrow's shape increases considerably the mating
call's efficiency and reach, and in order to build a burrow of the "correct" shape the
cricket actively "tunes" the burrow. The cricket digs a burrow, then tests its sound
quality by undertaking a test chirp. Based on the sound the cricket hears, it can
change the burrow shape and test the burrow's sound quality again, and so on, until
the burrow's shape converges on the exponential horn and the sound resonates
according to the cricket's preference. This process can take up to 2 h. As noted by
Turner, the singing burrow can therefore be seen as an extended phenotype *sensu*
Dawkins [71], a constructed niche *sensu* Odling-Smee et al. [276], or an extended
organism *sensu* him [359]. If the constructed niche persists across generations, it
can also be related to a kind of ecological inheritance that can be selected just as
genotypic inheritance would be. Lindholm's [232] article, "DNA Dispose, But
Subjects Decide—Learning and the Extended Synthesis", provides an updated
review of recent work on niche construction and further empirical evidence of its
occurrence in various groups. He supported the idea that organisms alter their
surroundings and affect the fitness of other species, thus commonly involving
elements of co-evolution, as Darwin recognized when he emphasized the importance of earthworms for soil development and vegetation. For instance, beaver
dams improve the environment for the beavers and allow them access to new

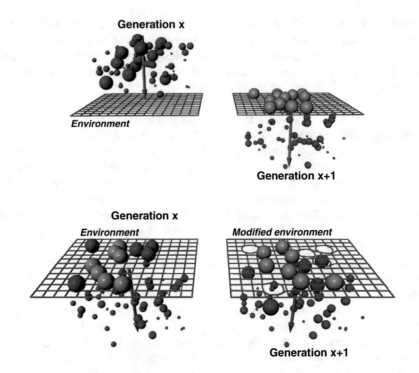

Fig. 3.1 The object-metaphor of the selective filter, according to, and modified from Turner [362]. In Neo-Darwinism (*left*), each generation contains within it variation in gene-objects that either pass through the selective filter to make up the next generation or do not. In niche-construction theory (*right*), certain biological agents can change the selective filter to allow themselves through

resources and protection from predators. Other examples of niche construction provided by Lindholm are the arboreal night beds of apes, the nests of birds that allow their offspring to grow up in "artificial" environments (oaks), and the anthills, wasp nests, or earthworm burrows.

The notion of niche construction thus plays a very important role in ONCE (Fig. 1.2) because it relates to the match between organisms and their "way of life" in the sense of behavior and ecology. That is, not in the sense of eco-morphological optimality, i.e. of a match between organisms and their habitat; rather because a major tenet of ONCE is the frequent occurrence of eco-morphological mismatches as explained in Chaps. 1 and 6. For instance, the fact that pandas continue—as a way of life due to behavioral persistence—to eat bamboo almost exclusively (and have a special "thumb" to help them do so) in a geographical region/habitat where bamboo has become more and more scarce (see, e.g., [294]) provides an example of both eco-morphological and etho-ecological mismatches (see also Chap. 5). It is interesting to note that the concept of "ecological inheritance" within the niche construction literature has similarities to the concept of behavioral persistence *sensu* Baldwin and *sensu* ONCE (Fig. 1.2). However, as will be explained later in the text, a main difference between ONCE and the view defended by some authors who write about niche construction, such as Turner [362], is that within ONCE there is no need, nor place, for the old and recurring teleological notions of evolutionary "purposefulness" and "intentionality".

Another point related to niche construction is that the term "organic" in ONCE refers to Baldwin's organic selection in general and not to the so-called Baldwin effect. In fact, the importance given to the Baldwin effect in recent decades is mainly due to the gene-centered view of evolution that is still dominant today [84]. As noted by Griffiths [161: 200], "when Baldwin describes how social heredity 'tends to set the direction of phylogenetic progress,' he does not describe a case in which a character initially transmitted by social heredity later becomes congenital (the Baldwin effect), but instead describes a case in which social heredity of one character changes the selection pressures on other, quite different characters of the organism (niche construction)." It is likely that in the first phases of the process there are no major genetic changes, or that the major changes relate for instance to modifications in the gene-expression profile. However, because both Baldwin's organic selection and ONCE incorporate the importance of random mutations and natural selection in later phases of the process (Table 2.1, Fig. 1.2), it is probable that some of the phenotypic traits, in particular anatomical ones, may be reflected in the genetic makeup of the species/population after several generations. Be that as it may, the crucial point is that ONCE does not require behavioral persistence to be related to specific changes in the genetic makeup/profile of a taxon by way of natural selection. As explained previously, behavioral/ecological inheritance is enough to explain the persistence of behaviors.

In fact, the more empirical data researchers accumulate about organisms and their behaviors, the more we learn that phenomena previously considered to be "innate instincts" fixed in the genotypes of organisms are actually further instances of cases of (non-genetic) social heredity associated with epigenetic factors directly influenced by the external environment, as noted by Moore [263: 124–125]. For instance, a newborn rat pup "approaches, attaches and sucks from the mother's nipple within minutes of birth; however if the nipple region is carefully cleaned of odors, the pup will not grasp the nipple and suck." Thus, "it seems that the pup uses odors that are naturally on the mother's ventral region to guide and elicit suckling behavior" and that these odors "come primarily from maternal saliva, which has odorants in common with those found in amniotic fluid." It has been experimentally demonstrated that the pups "learn about maternal odors in the uterus and use this learning to organize their neonatal suckling." In addition, it has been shown that even in the emblematic case of the newly hatched ducklings that direct their responses preferentially to maternal calls of their own species after isolation, "if the ducklings are both isolated and prevented from vocalizing (by transiently immo-bilizing the syrinx)", they do not exhibit this behavior. That is, "ducklings gain sufficient input from their own vocalizations to acquire preferential direction of filial responses to conspecific maternal calls." In fact, as reviewed by Sultan [350: 154], a careful overview "of long-term studies in vertebrates showed that most of the many reported adjustments to climate change probably were plastic responses and not, as reported, 'adaptations' based on genetic change."

Box—Details: Quantum   Biology,   Instinct,   Epigenetics,   Birds, Migrations, and Systems Biology

Interesting data are being obtained in the fascinating new field of quantum biology [251]. For instance, one of the most puzzling phenomena in biology concerns how organisms find their ways around the planet during long migrations. European robins (*Erithacus rubecula*) are one of the most emblematic cases. It is now known that these birds do not have any type of "map" imprinted in their genotype. Instead, they are able to detect the direction and strength of the earth's magnetic field with magnetoreception. That is, at least some of the mechanisms that enable the birds to do this might have been somehow assimilated in the genotype through genetic assimilation and/or the Baldwin effect because they would have been positively selected after the birds started to display migrating behaviors and to construct their own niche. However, the specific way in which the birds find their target is epigenetic and dynamic, in the sense that they have the ability to sense our planet's magnetic field and to then "draw" directional information from it by way of a navigational sense [251]. Without going into the details, it is now know that the nocturnal migrations of the robins require a small amount of light to activate their magnetic compass and that cryptochrome might provide the avian eye with such a quantum entaglement to navigate around the globe by way of the use of radical pairs necessary for fast triplet reactions. More details about this case study, and about the fascinating new field of quantum biology, are given in McFadden and Al-Khalili's 2014 book, *Life On The Edge—The Coming Of Age Of Quantum Biology*.

A deep implication of the data being obtained in this new field is that they are in line with some of the tenets of the also rising field of systems biology in the sense that the whole cannot simply be described as the sum of the parts. That is, these data support the crucial notion of emergence, which is also crucial for ONCE because it goes against a mechanistic, atomistic view of biology (see Chap. 1). As stressed by McFadden and Al-Khalili, at a microscopic level our smaller particles (electrons, protons, and so on) have dual particle-wave quantum characteristics, but surely our bodies as a whole do not display wave-like characteristics, behaving instead mainly as particles. For example, contrary to what you might have seen in some Hollywood movies, we unfortunately cannot cross concrete walls. A critical point of that book is also that the more we know, the more we see that senses are increasingly important to understand the biology and evolution of taxa and to explain features that were previously seen as purely fixed (genetic) "instincts". This reinforces the notion that, by using their senses, organisms are active agents that can sense, and thus take behavioral choices accordingly, i.e. they are not merely automata with fixed instincts coded by the genotype.

A field experiment reviewed by Scarfe [325: 263] further illustrates and supports Baldwin's notion of organic selection. In a 2006 study performed by Jonathan Losos involving ground-dwelling brown *Anolis* lizards on small islands in the Bahamas, changes in the morphologies of the lizards were seen over several generations after the controlled introduction of curly-tail lizards, a large predator, into their environment. Over the initial 6 months, brown anole lizards with longer legs were selected by natural selection because they had a faster running speed that assisted them to get away from the curly-tails. However, later many of the brown anole lizards seemed to adopt a new behavioral strategy (choice) that prevented predation more effectively: they became to be more arboreal and began perching on high branches where the bulky curly-tails could not reach them. Six months later, brown anole lizards with shorter legs, which better enabled them to climb and to reach their sanctuary, were selected for, thus demonstrating—as pointed out by Scarfe—"how a behavioral shift preceded a morphological change, the type of change that we most readily associate with the term evolution."

A more recent example concerns the variable behavioral breeding strategies of ambystomatid salamanders and specifically how empirical data suggest that variation in breeding ecology leads to divergence in embryonic development. According to Hale et al. [166], salamanders of the genus *Ambystoma* ancestrally were very likely aquatic with egg-laying and winter breeding, with early breeding having evolved in three species and terrestrial breeding in two of the three. Their results indicate that, because early terrestrial breeding is typically linked to expanded incubation, this type of breeding has led to selection for longer embryonic ontogenetic periods. Specifically, when they compared the embryonic development of the terrestrial breeding *Ambystoma opacum* with the aquatic breeding, co-occurring *Ambystoma maculatum* within dissimilar laboratory conditions over two breeding seasons, *A. opacum* embryos took longer to develop and hatched at a later stage. This result supported the predicted hypothesis that early terrestrial breeding favors embryos that have an extended development [166].

As noted in Chap. 1, since Baldwin's epoch, studies have provided hundreds of examples of behavioral shifts in non-human taxa that subsequently spread by imitation among a population. One of the most famous ones—potato washing in Japanese macaque—was widely discussed in the literature including a recent book that provides numerous other, less famous examples [216]. Many of these publications focus particularly on whether or not such examples can be defined as "culture". Personally, I am amazed to see how so many authors try so hard to discard all the potential examples of "animal culture" extensively described in the last decades, perhaps simply because they are still very reluctant to accept that one of the few remaining characteristics once thought to be unique to humans might also be taken from us, signaling the end of human exceptionalism (see, e.g., [86, 88–90]). However, this discussion is not central to the idea of ONCE. What is important is that these examples show how common behavioral shifts are in nature.

Neither is it important, in the context of ONCE, whether or not such behavioral shifts are necessarily related to "conscious" decisions or some particular kind of "intelligence", subjects that have been widely discussed by psychologists and ethologists including Baldwin himself (e.g. Bruce and Depew 2001). For instance, it is not needed to discuss in detail here whether positively phototactic organisms such as moths (e.g. Guhmann et al. 2015)—which move toward light sources—do so consciously. What is crucial for ONCE is, first, that they actively adopt this behavior when they see light sources (behavioral choice), meaning that they currently have a drive to do so, and, second, that they have done so over many generations (behavioral persistence *sensu* Baldwin), as explained in Chap. 1. One advantage of this way of thinking is that it does not create artificial divisions, which are often biased and in particular anthropocentric, between organisms that can be active players in Baldwin's organic selection *versus* organisms that cannot simply because they supposedly lack consciousness and intelligence. As the title of the recent provocative book by Frans De Wall [75] asks, *Are We Smart Enough to Know How Smart Animals Are?* The crucial point, therefore, is simply that organisms, including humans, bacteria, or plants, have a drive—through very simple or very complex mechanisms, it does not matter—leading to their behaviors and potentially to behavioral shifts and behavioral persistence.

This view was recently defended by Hoffmeyer [187: 157], who notes that "while the acceptance of agency as an inherent property of living systems does contradict the Darwinian ambition for establishing an externalist theory of evolution, it does not imply anthropomorphism in the narrow sense that ascribes specific human psychological properties to living systems in general." That is, "since we are descendants from earlier life-forms, it would be absurd to claim all of our properties to be unique for the human species… agency is a common property of life… as descendants from ancestor organisms that themselves possessed agency all the way back to the first organisms on earth." Or, as put by Kauffman [207: 6–7], "bacteria clearly" have "the capacity to make at least one discrimination, food or not food, and to 'act' upon that discrimination… and, without invoking consciousness, are therefore agents… the bacterium sense its world and act to avoid toxins and to obtain food." A similar idea was defended in Margulis and Sagan's book, *What is Life?*. In that book, they characterized evolution as the "sentient symphony" and claimed that the most significant aspect of the behavior of all living organisms is their ability to make choices, stating that "at even the most primordial level, living seems to entail sensation, choosing, mind" ([243: 180]; see also [61]).

As noted in Chap. 2, in the last few years, the view that any living organism can display behavioral choices has become more widespread in many areas of biology including the new field of biosemiotics, which is specifically based on an understanding of agency as a real property of organismal life [187]. In particular, data from many other biological fields, including neuroscience, are showing that many non-human organisms have more complex ways of thinking than was previously assumed. For instance, referring to vertebrates in particular, Lindholm [232: 338] stated: "through learning individuals begin to decide over themselves, and decision making depends on autonomous systems" that have "a physiological basis in the basal ganglia system". According to him, "even ancestral jawless fish such as lampreys comprise similar dual behavioral motor output systems like those of mammals, and their basal ganglia are fully developed…. lampreys are hence capable of regulating behavior by both direct and indirect motor output pathways from the basal ganglia, and possess the neurological basis for decision making."

Box—Detail: Do Insects, Crustaceans and Cephalopods Have Some Kind of Consciousness?

Barron and Klein [31] defended the idea that at least some insects may have some type of consciousness. This suggestion is not based on metaphysical speculations or theoretical divagations but on empirical data. Specifically, and using their own words, they proposed that at least one invertebrate clade, the insects, has a capacity for the most basic aspect of consciousness, i.e. subjective experience. In vertebrates this capacity is supported by a network of structures in the midbrain that create a neural simulation of the state of the mobile animal in space; such an integrated representation from the animal's perspective is sufficient for subjective experience. What they showed is that structures in the insect brain perform similar functions, suggesting that it also may support subjective experience. They argued that in both vertebrates and insects this form of behavioral control system evolved as an efficient solution to the basic problems of true navigation and sensory reference. For instance, the brain structures that support subjective experience in vertebrates and insects are very different from each other, but in both cases they are basal to each clade.

Within their article and the literature their review, a compelling case of a command function of the insect brain for the behavioral system of whole organism is the effect of injecting neurotransmitter agonists and antagonists into the central complex region of the insect brain. The authors pointed out that the parasitoid jewel wasp *Ampulex compressa* uses its ovipositor to inject venom containing GABA and octopamine antagonists into the central complex of its cockroach prey. The venom is not paralytic because the cockroach is still able to perform basic actions, but it disrupts the cockroach's behavioral program, thus rendering it passive, e.g. not struggling as the wasp leads it by the antennae into the wasp's burrow. The effect of *Ampulex* venom on the cockroach brain is such that it eliminates the ability of the cockroach to organize and initiate behavior, thus showing that the central brain structures are key for initiation and direction of movement in cockroaches and crickets.

While I was writing this book, several other papers and even whole monographs were published arguing that consciousness is in fact a feature that is much more broadly distributed among living organisms then was previously assumed. Within all of those publications, an emblematic example is the book *The Ancient Origins of Consciousness* (2016) by Todd Feinberg and my colleague Jon Mallatt. That book compiles a vast array of evidence from several fields of science and argues that all living vertebrates are conscious and that at least several extant invertebrates—such as insects, crustaceans (the clade including crabs) and cephalopods (the clade including octopuses)—meet many criteria for consciousness. In my opinion, the less biased and anthropocentric scientists will be, the more it will be realized that this list is probably still largely incomplete.

The article of Barron and Klein [31] mentioned in the previous box stressed that new functional analyses suggest that the insect brain supports a behavioral core control system that is functionally similar to that of the vertebrate midbrain. According to them, such analyses could help explain the dynamic and flexible behavior for which some insects are famous; for instance, some foraging ants and bees have astonishing navigational skills and spatial memory and are able to organize their behavior with respect to more than just their immediate sensory environment. This is because they execute targeted searches in suitable locations and at specific times for resources they have found previously. In addition, several insect species have been shown to plot new routes based on learned landmarks and goals. As the authors point out, the honey bee dance communication system requires a dance follower to determine a flight vector relative to celestial cues from symbolic and stereotyped dance movements. However, it should be noted, still using the authors' own words, that centralization of the nervous system is fundamental for subjective experience, but it is not sufficient. For instance, the nematode *Caenorhabditis elegans* has a simple and well-characterized centralized nervous system that can integrate input from an array of thermoreceptors, mechanoreceptors, chemoreceptors, and nociceptors along with a sense of time to organize directed movements. Nematodes are thus able to integrate multiple forms of sensory input using a centralized nervous system, but it seems that their behavior is organized as responses either to the immediate sensory environment or to immediate signals of physiological state. They argue that although nematodes do possess forms of memory that can change how they react to stimuli and take behavioral choices, there is no evidence that this memory has a spatial component or contributes to a structured model of their environment, as it does in mammals. This supports the idea emphasized in the present book: displaying behavioral choices, and undertaking behavioral shifts, does not necessarily requires the presence of consciousness or of particularly complex neurological skills. A last case study about invertebrates that stresses the importance of learning in these animals, and how learning contributes to the regulation of feeding, concerns locusts. They can link odors, color clues, and food locations with specific nutrients within foods, and can acquire associative food aversions to nutritionally deficient foods or toxic chemicals through post-ingestive mechanisms [337: p. 916]. As noted by Simoes et al., "this impressive array of learning mechanisms appears to be related to being generalists that must make decisions about which plants to feed on under competitive pressure; moreover, each food item presents a complex and variable mixture of nutrients and deterrent or toxic compounds, all of which require evaluation and may be informative for subsequent encounters".

Research on plants, particularly over the last few decades, has also supported this idea, by revealing that even plants—normally seen as the illustrative example of passive players, particularly because they supposedly usually cannot "move" and thus "make active choices"—also display an impressive array of active behavioral choices. Examples of such choices have been given in numerous journals including those specifically dedicated to this topic, such as "Plant Signaling and Behavior", as well as in many recent books. One of the books that I found particularly interesting and engaging is Mancuso and Viola's [241] *Brilliant Green: The Surprising History and Science of Plant Intelligence*. I should acknowledge that I do not appreciate the authors' repeated use of not only teleological but also anthropocentric terms in the book, although I understand that this is precisely what makes the book so engaging and provocative. Moreover, apart from being scientifically incorrect (in my opinion), the use of such terms can also be dangerous because it gives skeptics fuel to criticize the notion that non-human (or at least non-animal [for the less skeptical]) organisms can undertake any type of behavioral choices. For instance, on page 4 of the Introduction of that book the authors state: "plants talk to each other, recognize their kin… as in the animal kingdom, in the plant world some are opportunists, some are generous, some are honest, and some are manipulators, rewarding those that help and punishing those that would do them harm." One can easily understand how a skeptical scientist, or particularly a layperson, could easily use this example to dismiss the entire argument of that book, by pointing out, and correctly so, that for instance plants do not talk.

In my opinion, the use of such terms pays a disservice to an otherwise powerful and timely book with interesting and opportune historical notes about why plants were historically seen as such passive organisms as well as a well-documented, extensive review of several examples from empirical studies that clearly contradict this idea of behavioral passivity. Thus, as long as readers take into account that the use of teleological and anthropocentric terminology is more of a simplification/metaphor to convey an idea (e.g. plants "talking" refers to plants communicating with each other) based on scientific facts and empirical evidence, they will very likely appreciate the depth of the book. In particular, readers will gain a fascinating insight into the topic of plant behavior. Moreover, I strongly recommend that readers also refer to the original references cited in the book for further examples. These examples reveal a trend, within different fields of biology and among researchers studying all types of organisms, toward understanding that this diversity of behavior, and the crucial role played by behavioral choices, is actually not a rarity but rather a widespread commonality among all the organisms living on this planet.

Box—History: Aristotle, Religion, Linnaeus, Darwin, Plants, Souls, and Sleeping
In the beginning of their book, Mancuso and Viola note that the question of whether plants are "intelligent" dates back to Ancient Greece, when philosophers argued for and against the proposition that plants have a "soul". Aristotle divided living beings according to the presence or absence of soul, which, for

him, required "movement and sensation": the word animate itself means "having the ability to move". However, because Aristotle recognized that plants could reproduce, he opted for an intermediate solution: he attributed to them a "low-level soul". Religion in western culture also promulgated the idea that plants are "inferior" organisms as exemplified by the story that God gave instructions to Noah, before the Flood, to load onto the ark every creature that moved: plants were not included. Linnaeus was an exception among Westerners: he recognized that plants have "sex"—earning him condemnation for "immorality"—and even that they can sleep. However, one other feature of plants—that they can eat animals—was so incompatible with the ideas generally accepted in Linnaeus' period that he tried numerous other potential explanations rather than recognize it, as noted by Mancuso and Viola. For instance, he hypothesized that the insects did not die at all or that they actually chose to remain inside the plant.

The idea that plants sleep was renewed by Charles Darwin in the book *The Power of Movement in Plants* [68] and, for that matter, was just defended in a recent article by Puttonen et al. [300]. In another example of how Darwin was brilliantly ahead of his time, the book presented the results of several observations and experiments performed by Charles and his son Francis on the striking diversity of movements in plants. Mancuso and Viola spend a great deal of their book discussing this diversity, pointing out that the so-called "loss of movement" of plants is very likely a secondary trait. For instance, they refer to protists, such as the paramecium, and similar non-plant organisms that have chloroplasts as plants do, such as euglena. Significantly, organisms such as euglena can display behaviors typical of both plants and of animals. For instance, normally the euglena meets its energy needs by way of photosynthesis, as plants do, but if light is scarce it may transform itself into a predator and behave more like an animal by locating food and moving to reach it with the aid of very thin flagella. In fact, the euglena also has a rudimentary sense of sight that allows it to detect light frequencies and then to find the best position for receiving light as well.

As pointed out by Mancuso and Viola, modern techniques of photography and film have enabled us to document that plants do move to capture light, to distance themselves from danger, and to seek support, e.g. in the case of climbing plants. The positions of plant leaves are often re-oriented toward the light, a movement known as phototropism. Plants sense light by way of chemical molecules that act as photoreceptors and receive and pass on information about the direction from which light rays originate as well as about their quality. You might say: but that is not really an example of choice, because there is only one possible outcome, i.e. plants would always merely move toward the source of light like inanimate objects on earth move in only one possible way when confronted with gravity. But it is not so simple. For instance, some parts of a plant move toward the light, such as leaves

Fig. 3.2 Venus flytrap (*Dionaea muscipula*), a carnivorous plant (by Beatrice Murch [attribution: http://creativecommons.org/licenses/by-sa/2.0; https://commons.wikimedia.org/wiki/File%3AMeal_worm_in_venus_fly_trap.jpg])

(positive phototropism), whereas others move away from the light, such as roots (negative phototropism, or "photophobia"). So, different options do exist within the same plant.

In my opinion, the most powerful example of behavioral choice by plants is provided by carnivorous plants (Fig. 3.2). As explained in Mancuso and Viola and noted in the previous box, even authors such as Linnaeus—who argued in favor of the complexity of plants against the commonly accepted ideas of their time— refused to recognize the carnivorous nature of some plants. Such a behavior was simply unthinkable: how could an "inferior" plant be able to kill and eat a "superior" organism such as an animal? Many botanists tried to discount the plants' behavior by arguing that the leaves moved by reflex action and that the insects could have freed themselves if they had wanted to. However, such ideas could not explain many other aspects of the plants' behavior that became well known through empirical studies. For instance, plants such as the Venus flytrap (*Dionaea muscipula*: Fig. 3.2) never free the insects before killing and digesting them, and the leaves actually reopen soon after closing over something that is indigestible. In fact, many of the so-called "insectivorous plants" actually also eat small vertebrates such as lizards and even mammals such as mice, but until the mid-1800s the term "carnivorous plants" was still considered to be too strong for most biologists.

In short, the plants are not merely passive consumers of whatever touches their leaves. The leaves only remain closed for longer times when organisms of the very few specific animal species that make up part of their *chosen* behavioral repertoire —among the hundreds that can come in contact with the plants—touch them. When this happens, they do not let them go until the animals are killed and digested. According to Mancuso and Viola, the behavior of eating animals might be related to the humid marshes in which these plant species probably evolved millions of years ago, where nitrogen, which is essential for the life of plants (for protein production), was scarce or unavailable. In such nitrogen-poor habitats, eating animals would allow plants to obtain nitrogen that could not be obtained from the soil by their roots. How do plants such as the Venus flytrap catch their prey? The bait for the animals is a very fragrant, sugary secretion released on the leaf, which does not close when touched by any potential prey. If it did, it could catch something inedible or even allow an insect to eventually run away. Instead, the plant waits until a specific prey animal from its chosen repertoire lies in the middle of the leaf before closing it, because the prey activates the trap by touching at least two of the three small hairs situated in the region of the trap at an interval shorter than 20 seconds. When the animal is dead, the leaf starts releasing digestive enzymes to digest it.

Another example from Mancuso and Viola's book about plant behavior concerns the behavior of roots. A plant can have millions of roots that enter the ground and extend toward water and nutrients and/or away from dangerous substances. Importantly, if during this process a root meets an object, such as a stone, its growth is not interrupted, nor does the root change its route in a preset direction (e.g. always going downward or against the light). Laboratory experiments have shown that instead the root contacts the obstacles and then continues growing, twisting around the obstacle. This behavior is made possible by the extremity of the root, the so-called "root tip". Similarly, the aerial part of the plant, e.g. in climbing plants such as the pea vine, has multiple sensitive tendrils that curl up in a few seconds when they contact an object, normally twining around it. The last example I will provide about plants was reviewed by Sultan [350: 87]. Maternal plants in numerous clades actively place their offspring in appropriate germination sites in the soil—geocarpy—as seen for instance in the common peanut or groundnut *Arachis hypogaea*. As noted by Sultan, after fertilization "the small stalk on which the peanut's flower is born elongates into a specialized structure called a gynophore which uses light, gravity, and touch cues to grow downward, penetrate the soil, and bury the immature ovary underground", where "the seeds are protected from both desiccation and herbivores".

All of the phenomena described in these paragraphs—and the numerous others described in Mancuso and Viola's [241] and Sultan's [350] books and many other recent publications on plant behavior—thus cannot be reduced to mere "reflex actions". Nor are they comparable with events in the world of non-living objects. In fact, if this were simply a one-possible option or a single-automatic option, plants would not display the striking evolvability they do. They would not change drastically over time as they do (in contrast to non-living objects). Plants change not only their phenotypes but also their prey repertoires: insects only in some plants, lizards in others, other reptiles still in others, and/or even mice in some. Therefore, the behavioral choices of plants—and, again, I do not make any statement about consciousness—did change over time through evolution. To reject this conclusion and deny that plants display complex behaviors and behavior choices in general would be to follow an historical bias that has influenced biological and philosophical sciences for far too long.

Box—Detail: Plants, Insects, and the Animal-Centered Ideas of Most Biologists

One of the aspects of Mancuso and Viola's stimulating book that I deeply appreciate is that it questions not only our human-centered but also our animal-centered view of evolution. In addition to the example discussed previously regarding the scientific community's historical difficulty in accepting that plants can not only catch but also kill and digest animals, they provide other examples of encounters between plants and animals in which plants clearly seem to be the ones "in control". As most readers probably know, flowers of the orchid *Orphyrs apifera* can mimic the shape and consistency of the tissues and the scent of the female of certain nonsocial hymenoptera (i.e. insects that are similar to wasps and bees but that do not live in colonies). The male insects are highly attracted by these flowers and copulate with them, often even preferring them to females of their own species. When the male insects start copulating with what they presume to be the females of their species, the flower showers their heads with pollen, which the insects take with them when they go to, and pollinate, the next flower. Even if one thinks that the insects have some type of ecological/evolutionary gain in such interactions—and adaptationists, particularly animal-centered ones, could probably list hundreds of possible theoretical gains—Mancuso and Viola's [241: 114] statement about these interactions at least makes one reflect on their meaning: "just which one is running this show, the plant or the insect, seems absolutely clear." As does their argument that, if we discovered a faraway planet that is 99% inhabited by a certain form of life, we would very likely say that the planet is dominated by that life form. According to the authors, 99.7% of the earth's biomass (the total mass of all living organisms) is made up of plants, whereas humans together with all other animals represent only approximately 0.3%. Therefore, why would we say that our planet is dominated by humans, and not by plants?

I have already argued in the previous chapters that unicellular organisms, such as bacteria, also display behavioral choices, and I will give further evidence to support this argument in the chapters below. Here I mainly provide an illustrative example among the many other presented by Sultan [350]. As reviewed by her, experiments with the marine bacterium *Pseudoalteromonas haloplanktis* revealed that rapid swimming responses to chemical stimulus allowed the bacteria to exploit temporary resource patches. For instance, in response to lysed algal cells, the bacteria clustered within tens of seconds in foraging "hot spots" that had three times more bacterial cells than other microsites. This is clearly a case of organismal behavior, in which whole organisms have the possibility and capacity to choose/decide and have the drives to undertake those choices or not. This is attested by the fact that there is no single possible, "automatic" outcome in which bacteria basically behave such as automata or as rocks falling under a similar constant acceleration due to the force of gravity. Instead, the outcomes are clearly different, with for example the fastest-moving 20% of the population experiencing a tenfold higher nutrient environment than non-motile cells.

Returning now to the topic of consciousness, and taking into account the dangers of overusing this term, I would like to stress that there are many cases in which behavioral choices/shifts could be in fact considered to be the result of a conscious decision, but that ultimately lead to outcomes that were not predicted consciously. An example, which will be discussed in Chap. 5, concerns the lighter skin of humans who live in Northern regions such as Norway or Iceland. The very first *Homo sapiens* individuals probably made a conscious choice to begin the long-term migration from their previous home regions to other areas because of environmental stress and/or to seek better conditions. However, the humans who started the trip did not point to a map—maps did not even exist yet, obviously—showing "Norway". They probably did not even decide consciously to go to Northern cold regions before starting their migrations. In fact, it was not those individuals who initially chose to migrate but their progeny, many generations later, who arrived and colonized those cold regions. Therefore, there is nothing deterministic, finalistic, when one argues that living organisms can, and do, take behavioral choices, or in any one of the related ideas that make up part of ONCE.

Similarly, counter to the narrative that organisms are mainly passive players in phenomena such as metamorphosis, numerous examples now show that an initial behavioral choice/shift—e.g. tadpoles of the New Mexico spadefoot toad feeding voraciously on their crustacean diet, which is high in thyroid hormone—must take place first, subsequently leading to metamorphosis [171]. However, as pointed out in Chap. 1, I would not invoke any type of vitalistic or teleological notion of purpose or goal-directed process assuming that when the tadpoles choose (consciously or not, it does not matter) to feed voraciously on their crustacean diet, they —or any "higher entity", be it God or Mother Nature—consciously knew that this choice would lead to the subsequent metamorphosis.

Box—Detail: Metamorphosis, Memory, and Behavioral Changes

Another interesting question related to metamorphosis and behavior, as well as to mental capacities such as memory, is whether or not memory is retained through metamorphosis. Can a moth remember what it learned as a caterpillar? This question was addressed in an experimental work in which fifth-instar *Manduca sexta* caterpillars received an electrical shock that was associated with a specific odor to create a conditioned odor aversion and were later assayed for learning in a Y-choice apparatus as larvae and then as adult moths [39]. As expected, the larvae learned to avoid the training odor. Notably, this aversion was also still present in the adults and was not the result of leftover chemicals from the larval environment because neither applying odorants to naïve pupae nor washing the pupae of trained caterpillars resulted in a behavioral change. As the authors pointed out, these results have major ecological and evolutionary implications because retention of memory through metamorphosis can influence for instance behavioral choices related to host choice by insects and shape habitat selection, therefore eventually leading to sympatric speciation and micro-evolutionary and potentially also macro-evolutionary changes.

Before passing to the next chapter, I would like to refer to other examples of organic selection that have direct implications for the understanding of evolutionary trends toward a phenomenon that, in my opinion, is not discussed as often as it should be in the evolutionary literature: asymmetry. A particularly enlightening review on this topic was given in Neufeld and Palmer's [269] chapter "Learning, Developmental Plasticity, and the Rate of Morphological evolution", in Hallgrimsson and Hall's book *Epigenetics: Linking Genotype And Phenotype In Development And Evolution*. Their examples come from theoretical models as well as empirical data obtained from their experiments and those of other authors. For instance, they refer to experiments showing how, when American lobsters (*Homarus americanus*) are allowed to make their own behavioral choices to use both claws freely to manipulate hard objects or to interact with another juvenile lobsters during a critical period of the fifth molt, whichever claw is used most transforms into the crusher claw. Autotomy or denervation of the cleft claw almost exclusively induces the crusher claw to develop on the right side, whereas increased exercise of the left claw induces development of the crusher claw on the left side. The developmental program related to the production of a crusher claw thus depends on handed behavior to initiate it. This is an illustrative example of function (behavior) coming before form not only during macroevolution—Neufeld and Palmer state that handed behavior normally precedes handed morphology phylogenetically—but also during development. In other cases, such as that of paired penis-bearing earwigs (*Labidura riparia*), asymmetry appears to be mainly behavioral. Ninety percent of these earwigs hold the right penis in the "ready" position despite the fact that there are no apparent morphological differences between right and left penises and that the left penis is fully functional.

These cases can be seen as examples of Baldwin's organic selection without the so-called "Baldwin effect". That is, there seems to be no effect on the genetic makeup through natural selection because the asymmetry is exclusively, or at least mainly, due to behavioral persistence: most progenitors show/teach their descendants and/or the latter imitate/learn to use a certain feature of one side of the body more often. Considering the importance of asymmetry in evolution, including macroevolution, these examples further support the idea that Baldwin's organic selection and, in particular, the behavioral choice/persistence of organisms plays a central role in biological evolution. More examples and a deeper discussion on this topic are given in Neufeld and Palmer's [269] chapter and references therein as well as in books such as Pigliucci's *Phenotypic Plasticity—Beyond Nature and Nurture* (2001) and West-Eberhard's *Developmental Plasticity and Evolution* (2003)—as well as in more specialized monographs such as Malashichev and Rogers *Behavioral and Morphological Asymmetries in Amphibians and Reptiles* (2002). I will also provide more examples and discuss these subjects in further detail in the chapters below.

Chapter 4
Evolutionary Trends, Sexual Selection, Gene Loss, Mass Extinctions, "Progress", and Behavioral Versus Ecological Inheritance and Novelties Versus Stability

When confronted with the increasing number of publications reviving Baldwin's organic selection, many authors—particularly Neo-Darwinists—raise the following question: why do we need organic selection if Neo-Darwinism and natural selection are enough to explain the evolutionary phenomena seen in nature? Or, as stated by Downes [115: 34]: which phenomena seen in evolution call for special explanations not provided by Neo-Darwinism? I addressed this question to some extent in Chap. 1. A common answer given by Baldwin's supporters is that organic selection can be particularly useful in explaining cases of very rapid evolutionary changes. I agree—and this is a major point of this book—that organic selection *sensu* Baldwin can be more powerful than (external) natural selection *sensu* Neo-Darwinism. This is because organic evolution is mainly driven by the organisms themselves rather than by random mutations that may eventually lead, or not, to phenotypic changes that in turn may eventually be advantageous, or not, in a certain habitat that the organisms may, or may not, occupy.

In addition to the many other examples provided in this book, an illustrative example of the power of organic selection was given by Olson [278: 281]: "artificial selection"—i.e. a subset of organic selection mainly driven by humans—in many opposing directions has resulted in a diversity of skull shape in domestic dogs that far exceeds that in wild canids. Of course, there is nothing artificial in "artificial selection". It is simply part of Baldwin's organic selection but made by humans instead of by other organisms. It is one more example of the tendency to think that humans and their evolution are different from anything else seen in nature. Darwin himself emphasized repeatedly the unparalleled power of what Baldwin later designated as organic selection, in not only "artificial selection" but also in another type of selection that is also clearly driven by organisms and not by the "external" environment: that of sexual selection. In modern terms, one would say that both "artificial" selection and sexual selection and other types of Baldwin's organic selection are particularly powerful because they can explore, in a much more direct way, both the hidden and non-hidden plasticity of organisms than external (natural) selection usually can. A crucial source of the power of organic selection is

R. Diogo, *Evolution Driven by Organismal Behavior*,
DOI 10.1007/978-3-319-47581-3_4

behavioral persistence by the organisms either of the organisms being selected or those doing the selecting. For example, humans are consistently intervening in the selection of certain breeds/characteristics of dogs over many generations, so there is an "enduring persistence" *sensu* McShea, compared with Darwinian external (natural) selection (see Chap. 1) because of the highly variable and random characteristics of the external environment.

The huge evolutionary potential of sexual selection is well illustrated by numerous empirical case studies reviewed by Jennions and Kokko [201: 344]. They argue that the power of sexual (organic) selection followed by natural selection is particularly impressive when one takes into account that a gene for a sex-specific trait is only available to direct selection when it is expressed in the appropriate sex. For example, half of the genes for a favored male trait are hidden from selection because they are in the body of a female. Moreover, when these genes are expressed in females they are actively selected against if they produce a so-called "inferior phenotype". However, according to these authors sexual selection is still remarkably powerful because, unlike much of natural (external) selection, it is based on a zero-sum game in which one organism's win is another's loss. For instance, when there is competition among males, if female reproduction limits population growth, then for every offspring produced by a certain male, another male loses out on an opportunity to reproduce. That is, there is no single goal or common end point to sexual selection, contrasting with natural selection, which often selects for a single target.

Having said this, I do not believe that the speed of evolution is the main problem for Neo-Darwinism. Neo-Darwinists have ways of explaining cases of fast evolution by phenomena such as directional genetic selection, strong short-term climatic selective pressures, and certain cases of genetic drift, which can often be linked to each other. As explained in Chap. 1, in my view the two major problem of Neo-Darwinism are instead the frequent occurrence of long-term evolutionary trends and of etho-eco-morphological mismatches. The difficulty of Neo-Darwinists in explaining macroevolutionary trends was deeply appreciated by Baldwin: he explicitly used this difficulty to defend his organic selection theory using a specific term to refer to the directional influence of organic selection on evolution: orthoplasy [64]. In contrast, the widespread occurrence of etho-eco-morphological mismatches was mainly neglected in Baldwin's epoch (Chap. 1), due to historical biases but also to a lack of proper phylogenetic methodology (see also Chap. 6), and thus was not taken into account in Baldwin's works.

However, there is now vast empirical evidence that sexual selection can lead to evolutionary trends and etho-ecological and/or eco-morphological mismatches. Regarding trends, Fricke et al. [134: 406] reviewed work on pond skaters where there is sexual conflict because males gain from higher mating frequencies and have a wide variance in mating success, whereas for females mating is costly in terms of reduced foraging time and higher predation risk. Males are selected to try to mate with all available females, and females usually attempt to resist redundant matings, thus resulting in the occurrence of violent premating struggles in which the females attempt to avoid courting males. Importantly, experiments show that males with longer abdominal claspers, which aid them to attach to females during mating, are more likely to have higher mating success, whereas abdominal spines in females, which aid them in lowering mating frequency, lead to shorter premating struggles and a lower mating rate. That is, this sexual conflict has seemingly lead to antagonistic selection that in turn results in an evolutionary trend where males tend to have longer claspers and females tend to display longer spines. As stressed by Fricke et al., the idea that the armaments in males and in females are coevolving—due to behavioral persistence and directional selection as argued in ONCE—across different species of pond skaters is supported by both experimental and comparative evidence. In addition, there is ample empirical evidence that behavior and organic selection—including sexual preferences and selection—play a major role in speciation events [216].

Wiens et al. [402: 2082–2083] also provide an insightful empirical study that illustrates how sexual selection can lead to both long-term macroevolutionary trends and etho-ecological and/or eco-morphological mismatches. They used a phylogenetic approach to analyze the evolution of dorsal crests in European newts (semiaquatic amphibians of the urodele family Salamandridae), a well-known sexually selected character system that was first noted by Darwin. Their phylogenetic results show a general relationship between the evolution of some male display behaviors and the evolution of crests as well as that the collection of novel elements of the behavioral displays is related to the accumulation of modifications of the crests. They argue that the correlated addition of new elements to both the morphological and behavioral displays might be seen as a trend toward increasing complexity in both signal types. Their results also suggest that phenotypically plastic traits, such as the crests, can be maintained for relatively long macroevolutionary timescales, i.e. for dozens of millions of years. Therefore, those results completely fit the predictions of ONCE that organic selection *sensu* Baldwin (in this case sexual selection), mainly driven by the behavioral choice of the organisms (here, the preference of females for males with dorsal crests), can lead to long-term macroevolutionary trends. This occurs by way of behavioral persistence of the organisms associated with natural selection favoring genetic/epigenetic changes that are advantageous in the context of that behavior/niche (here, the morphological accumulation of crest modifications).

Moreover, as also predicted by ONCE, their results also show that whereas in some cases of long-term trends the plasticity seems to be lost, and both the behavior (by way of behavioral persistence) and form persist, in other cases the plasticity is not decreased. For instance, there is still enough plasticity for either behavioral shifts or morphological shifts, which can potentially lead to etho-morphological and thus likely to etho-ecological and/or eco-morphological mismatches. That is, the behavior might change, whereas the form does not: females may no longer prefer males with crests, but crests continue to be present. Or form can change but the behavior not: females may continue to prefer males with crests, but for some reason crests are lost. Of course, the two shifts can be combined: females may lose interest in males with crests, and crests are lost, but this is not a case of etho-ecological mismatch. The best-supported model of Wiens et al. [402: 2082–2083] indicates that dorsal crests evolved once and then were lost five times within newts. They note that repeated loss of sexually selected morphological traits suggests that sexual selection favoring such traits might have been weakened (as suggested by empirical studies finding that loss of these traits may be associated with the loss of female preference for those traits), that natural selection against them might have become stronger (e.g. related to other behavioral shifts in habitat use), or a combination of both. They state that in at least one genus included in their study—*Lissotriton* —"smaller size may make males more vulnerable to predation, possibly causing the costs of crests associated with natural selection to outweigh their benefits due to sexual selection." This would be a case of change in morphology without a change in behavior, which would thus differ from most of the other cases found in the literature reviewed for the writing of the present book (see Chaps. 1 and 5). Because I will focus on mismatches in Chap. 6, I refer readers to that chapter and to other examples of mismatches given in works that focus specifically on sexual selection such as Jennions and Kokko's [201] chapter in the book *Evolutionary Behavioral Ecology*.

The intricacies involved in mate choice also emphasize a point that is key to ONCE: the complexity of certain behaviors and the almost endless number of possible behavioral combinations further challenges the idea that organisms are mainly passive players or automata that simply follow a 'genetic or developmental' program. Brooks and Griffith [47: 416] provide a series of examples that illustrate this point well in their chapter "Mate Choice". For instance, in crickets there are normally various stages in a successful mating during which females have the opportunity to exercise mate choice. Male field crickets rapidly stridulate their forewings to broadcast a loud advertisement for females. Therefore, the first way in which females choose is by locating males on the basis of this call, and they do so by actively walking up the gradient of sound toward the male. Second, females are only able to localize males that display a continuous call for a certain minimum period of time. Therefore, females must hear the calls for a certain duration and thus are more likely to actively find the males that not only have a call structure appealing to them but also call for the longest durations and on most nights. Then, when a female has actively found a male, he stops his advertisement call and starts courting her with another type of call—courtship calls—which are quieter than advertisement calls. This call is needed for a female to decide to finally mount a male for mating. Experimental studies show how the properties of the call influence the latency for mating and how females seem to be able to exercise choice in the field at this stage by breaking off the interaction on the basis of those properties [47].

In addition, females also exercise choice immediately after mating by removing (or not) the externally attached spermatophore at some point after copulation, thus interrupting insemination. As noted by these authors, females appear to favor, for instance, males that are more attractive in courtship or larger males by leaving the spermatophores of these males attached for longer periods of time. Furthermore, the complexity of the entire sexual process is augmented by the fact that females can actively choose whether or not to mate with more than one male and/or more than one time with the same male. For instance, females can recognize and avoid re-mating with previous mates. Therefore, the females, and obviously also the males, play highly complex and active roles in the entire process. This example shows how a behavioral shift at any of the various stages of the process could lead to new behaviors and easily could potentially lead, for instance, to speciation due to reproductive isolation. It also shows how long-term behavioral persistence at all stages of such a complex process can lead to evolutionary trends in the evolution of a wide range of traits. Such traits might be related to the size/adherence of spermatophores; the forewings and their capacity to display loud advertisement calls, or other structures related to the display of courtship calls and/or the sensory and neuronal processing capacities of the female to hear/process the calls of the males [47].

As I will further refer to sexual selection in Chap. 5 in connection with arguments about Lamarckism, I will now focus more on the topic of evolutionary trends. It has been recognized for more than a century that the Neo-Darwinist view of evolution (Table 2.1) has major difficulties explaining the common occurrence of long-term macroevolutionary trends [265]. For instance, Johannsen [202] argued that natural selection has very limited power to drive long-term directional evolutionary changes. As shown in his famous 1911 study, and in similar laboratory studies since then, one can select some more or less extreme phenotypes associated with random mutations in inbred laboratory lines for several generations, but still nobody has really observed cases of well-defined, long-term evolutionary trends in a laboratory thus far. In my opinion, one of the reasons why these studies failed to reproduce such trends is because the researchers focused mainly on genetic changes, not on the behavioral shifts/choices of the organisms themselves. Therefore, a distinction must be made between the phenomenon designated "directional selection" by many Neo-Darwinists and macroevolutionary trends.

As explained previously, Neo-Darwinism has no problem explaining directional selection as defined in, for instance, population genetics. According to them, directional selection is a mode of natural selection in which an extreme phenotype is favored over others, thus causing the allele frequency to shift in the direction of that phenotype over time as a consequence of differences in survival and reproduction among the different phenotypes [262]. Because under this mode the main cause of directional selection is different and changing environmental pressures, rapidly changing environments—such as climate change—can in fact produce dramatic changes within populations. However, the problem is that this means that directional selection normally would not be sustained over long geological periods [186]. Directional selection *sensu* population geneticists thus is not expected to lead to major macroevolutionary trends, which refer to long-term directional changes within a single lineage or to homoplastic (convergent or parallel) changes across various lineages.

A major problem related with Neo-Darwinism is the occurrence of macroevolutionary morphological novelties such as tetrapod limbs or the mammalian inner ear. I include this difficulty within the major problem of explaining evolutionary trends because such novelties are often the outcome of long-term evolutionary trends (see, e.g., [245]). Basically, natural selection can only select among behaviors/anatomies that already exist and thus will always lead to reducing the existing behavioral/anatomical diversity overall as pointed out by Baldwin (Chap. 2). Unless driven by other phenomena, natural (external) selection will not normally lead to an increase in diversity or, even less likely, to the origin of new and complex novelties, particularly over long geological times in which the external environmental conditions will often change repeatedly. As Matsuda [245] stated, there is empirical evidence that most macroevolutionary novelties often occur in stressful environments, which often are associated with behavioral shifts by the organisms that live there. A clear example provided by the author concerns the links between behavioral shifts of salamanders in stressful aquatic environments and changes in epigenetic factors such as endocrine activity, which, together with other factors, may lead to neoteny, a phenomenon often associated to major evolutionary changes and trends (see Chaps. 7 and 8).

These issues were discussed in Stanley's [347] *Macroevolution* book. He stated that most macroevolutionary trends do represent changes within branched phylogenies—as clearly exemplified by the clade including horses—rather than evolution within single lineages, so in this sense they mostly involve cladogenesis, rather than anagenesis. He then linked behavioral complexity with variability, cladogenesis and evolvability, and therefore with macroevolution, by stating that empirical data show that more behavioral complexity is usually related to a shorter time of duration of species. This is because, according to him, more behavioral complexity leads to a greater potential to undertake more behavioral choices, which in turn facilitate new directions of evolution and thus speciation, in which an ancestral species ends its existence because it forms—via cladogenesis—two or more derived species. However, Stanley's book illustrates the problem of most macroevolutionists to deal with, or explain, evolutionary trends. This is because if behavior is so flexible, and behavior complexity moreover leads to more frequent speciation events and the rise of new species, one could argue that macroevolution would be rather unstable and chaotic, while we know that in fact it is actually rather stable, with some taxa, including individual species, existing for millions and millions of years. That is, most macroevolutionists could not discern potential 'factors or persistence' that could explain macroevolutionary, including behavioral, stability.

In particular, they failed to recognize the importance of a crucial factor pointed out by Baldwin, and defended in the present book: *behavioral persistence*. Developmental constraints and behavioral persistence are powerful factors leading to a rather stable macroevolution. As already explained in this book, all organisms seem to have the *ability* to undertake behavioral choices and, in a few cases, embarking on unprecedented choices may lead to evolutionary advantages that will in turn result in behavioral shifts and the rise of innovations and/or new, successful evolutionary lineages. However, in most cases unprecedented choices do not lead to such outcomes. For instance, they may result in tragic consequences for the organism(s) doing them, as they did in the case of the lost bee that I found dead in my bathroom (see later text). Or they can lead to the discovery of a new favorable niche/condition, but because they are only done by one organism, or very few isolated organisms, this outcome may not lead, in the long term, to a major shift within the ancestral species and/or to the formation of a new species. It is the eternal red thin line—and balance—between not risking enough *versus* risking too much, similar to the balance, at a molecular level, of a system that is too inflexible to be mutated and thus cannot evolve *versus* a system that is too easily mutated and thus can quickly evolve but also lead to chaos and/or extinction.

Some readers could still argue: so, what happens in species that are not social, and particularly in which the young never see their parents, as for instance in aquatic species that lay eggs that are immediately dispersed to far way regions? How can one explain behavioral persistence, and thus macroevolutionary behavioral stability, in these cases? The answer, in my opinion, has mainly to do with the difference between ecological inheritance and behavioral inheritance (Figs. 1.2 and 1.3). That is, behavioral inheritance *sensu* current Evo-Devo is mostly related to phenomena such as teaching, learning and/or imitation, often between the young and their parents and/or other members of their populations, being therefore essentially similar to Baldwin's social heredity. Ecological inheritance is instead mainly related to the stability of the ecological networks between different populations and/or species and the other biotic, as well as abiotic, characteristics of the habitats that they inhabit. Therefore, both ecological and behavioral inheritance are linked to behavioral persistence. For instance, pandas only continue to display the behavioral persistence to eat bamboo since a long time ago because bamboo has been present in the areas they inhabit or in nearby areas since they started to display that persistence. If bamboo was not part of the ecological network that also includes pandas, pandas would either get extinct or be forced to eat other food items. Similarly, even in the case of young individuals that never encounter their parents, they will often inhabit, *by inertia*—i.e. if they to not behaviorally choose to do otherwise—similar habitats as those inhabited by their parents, e.g. because the currents will disperse the eggs to the same areas to which they dispersed the eggs of the previous generation, and so on. So, when an organism develops in those areas, probably they can only start feeding from one of very few available preys and/or other food items, and moreover their morphology—particularly due to developmental constraints—is more appropriate to eat only certain of those ecologically available preys and/or other food items. As a different, but related, example, if an Inuit that would be abandoned or lost since an early age would be able to survive in the harsh cold conditions that Inuits normally face, very likely it would be because he/she ate the same - very limited - food items that their parents ate, or at least similar types of items. Moreover, in many cases even if organisms are not necessarily social, or never saw their parents, they often still encounter, at least occasionally, other individuals of the same population or species, which they can for instance imitate. So, both ecological and behavioral persistence are in fact strong factors that can contribute to behavioral persistence and thus to macroevolutionary stability, particularly when the two are combined. Viable behavioral shifts are therefore relatively rare, as viable mutations are, but that is precisely why they are so significant and important for the origin of evolutionary innovations and for both microevolution and macroevolution, as explained in, and exemplified by numerous empirical cases presented throughout, the present book.

A widely discussed and beautifully illustrated example of a both a macroevolutionary trend and the stability of macroevolution is shown in Fig. 4.7. Osborn [287] reconstructed the skulls of different fossil taxa of the family Titanotheriidae—

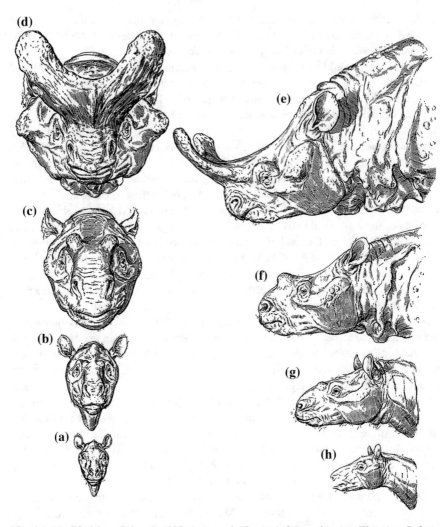

Fig. 4.1 Modified from Osborn's 1929 monograph *The titanotheres of ancient Wyoming, Dakota, and Nebraska.* Bottom to top—that is, phylogenetically basal to derived—are reconstructions of the titanothere genera *Eotitanops, Manteoceras, Protitanotherium,* and *Brontotherium,* which range from approximately 55 (A and H) to 35 (D and E) million years

also designated Brontotheriidae, ranging from approximately 55 to 35 million years ago, from the order Perissodactyla that also includes rhinoceroses and horses and—and argued that the evolutionary changes in the size of their horns were not random. According to him, these changes were instead biased in the direction of increasing horn size because various different titanothere lineages experienced the same type of change in horn size. Osborn, a Neo-Lamarckian, defended that the horns were increasingly useful for fighting but that the initial bumps were not functional, and thus that these directional evolutionary changes could not have been the result of natural selection [35]. That is, the directional changes must have been due to some direction in the process of variation itself, which assured the same evolutionary outcome time after time, although Osborn did not specify how such a process could occur, stating that this might possibly even be "beyond human solution" [35]. Neo-Darwinists therefore used this example to criticize Neo-Lamarckians by arguing that they were coming back to old vitalist ideas. They argued, for instance, that even rudimentary horns are better than nothing for fighting and therefore that the directional changes might have been simply the outcome of natural selection of chance variations [35]. Regarding the trend itself, they postulated that it could be simply a byproduct of selection for increasing body size and/or the result of selection on horn size directly (e.g. by selecting males that tend to win fights or by sexual selection by the females: see [35, 169]).

> **Box—History: Osborn's Orthogenesis and the Recurring Discussions on Design, Purpose, and Vitalism**
> As explained in Chap. 1, discussions on evolutionary trends tend to be related to broader discussions on teleological topics such as the notion of "design", "purpose" and "progress" in evolution, which are also linked to issues such as complexity and recurring theme in biology since Aristotle to the present days. As stated just previously, Neo-Darwinists criticized Neo-Lamarckians such as Osborn by coming back to old—and also recurring—vitalist ideas. In fact, as noted by Ruse [318], Osborn's main orthogenesis idea did relate evolutionary trends to a notion of design/purpose because he saw progress in the evolutionary history of organisms toward an increasing perfection of form, function, and beauty. In a recent paper, my colleagues and I discussed the notions of scala naturae and evolutionary progress, provided references to several classic and also more recent papers and books on the subject, and explained why we disagree with both of those notions [109].

As noted in the previous box, ideas such as those of Osborn, and in particular discussions on evolutionary trends, are often related to debates on whether there is a scala naturae and/or a trend toward 'progress' in evolution, both of which are contradicted in studies performed by my colleagues and me, particularly concerning phenotypic evolution in vertebrate/human evolution (e.g., [86, 88, 109]). A recent review by Albalat and Canestro [5] also contradicts the existence of a general trend or 'progress' toward more complexity at the genetic level. They note that great attention has been paid to the mechanisms of evolution by gene duplication, whereas gene loss has often been associated with the loss of redundant gene duplicates without obvious functional consequences and therefore is mostly neglected as an evolutionary force. However, new genomic data reveal a new perspective of gene loss as a pervasive source of genetic change that has great potential to cause adaptive phenotypic diversity. In particular, studies on comparative genomics showed that the ancestral eumetazoan genome was much more complex than expected and that gene loss was pervasive in many animal phyla. As they noted, this new perspective challenges the traditional notion that had influenced the analyses of the first known genomes (the nematode *Caenorhabditis elegans*, the fly *Drosophila melanogaster*, the plant *Arabidopsis thaliana*, and *Homo sapiens*): the scala naturae, in which attempts were made to correlate the supposedly increase of biological complexity leading up to humans with an increase in the number of genes.

According to this new perspective, Albalat and Canestro refer to the "less is more" idea: non-functionalization represents a frequent evolutionary adaptive response that may be relevant when populations are exposed to changes in the patterns of selective pressures due to major environmental changes. They review a series of empirical studies showing how adaptive gene loss has been crucial in unicellular organisms, e.g. in bacteria more than 200 examples of gene loss have been linked with adaptations to environmental shifts. Meta-analyses of bacterial genome-wide fitness data (from transposon insertion and inframe deletion mutations) across 144 conditions reveals that adaptive null mutations are strikingly abundant and disproportionately affect both enzymatic and regulatory pathways. In addition, a null mutation could be adaptive in many different conditions, and numerous cases of gene loss in bacteria have been associated with adaptive gains in pathogenicity. For instance, in the pathogen *Candida glabrata*—infamously known for being a causative agent of vaginal infections—the loss of de novo biosynthesis of nicotinic acid genes has been positively selected as a mechanism increasing pathogenicity by way of direct infection to the urinary tract in experimental mice. As also stressed by Albalat and Canestro, some of the most well-known empirical examples supporting the "less is more" idea were described in humans. For example, loss of function mutations of the C–C chemokine receptor type 5 and atypical chemokine receptor 1 provide resistance to AIDS and vivax malaria.

Linked to gene loss is another topic that, as noted in Chap. 1, is crucial for ONCE (Fig. 1.2): the importance of randomness in evolution, which is linked to phenomena such as genetic drift. As Albalat and Canestro [5] state, examples of gene loss associated with regressive evolution could be considered *a priori* to be evolutionarily neutral because each loss can be related with the loss of a biological feature that seems expendable under new environmental conditions. However, it could be argued that the loss of genes is under positive selection due to the advantages that it might provide (e.g. energy savings and/or spatial efficiency). According to them, comparative genomic analyses in bacteria suggest that the decrease of genome size is mainly driven by genetic drift and that there is no evidence to support a link between smaller cell size and environment nor is there any selective advantage of smaller genomes due to a decreased metabolic burden of replicating DNA. Such a random character of neutral evolution would thus help to explain examples of eco-morphological mismatches, which are so important in ONCE. In addition, neutral evolution could also be key in cases of eco-morphological matches such as the loss of eyes in cavefish. As they point out, there is ongoing debate on whether the loss of cavefish eyes is neutral—as indicated by the independent accumulation of various non-functional mutations in multiple eye–related loci—or is an indirect consequence of a hitchhiking effect promoted by positive selection on a nearby locus. I will come back to this subject and to cavefish later in this chapter.

Similarly, and also related to understanding the evolutionary dynamics of gene loss, is the question of whether most gene loss fixations, per se, are neutral or adaptive. As explained by Albalat and Canestro, this question concerns the wider and still open neutralism–selectionism discussion on whether genetic variation found in populations is mainly neutral or adaptive and whether neutral variants may be important to the origin of evolutionary novelties. These authors point out that according to the tenets of population genetics, the probability of fixation of neutral gene losses depends exclusively on the population size, and thus on genetic drift, whereas in adaptive gene losses the fixation probability is also dependent on the selective coefficient. Therefore, even if a gene loss is slightly deleterious, it has a small probability of becoming fixed in the population by drift. According to their review, for free–living unicellular organisms with large population sizes, gene-loss fixation seems to have been mainly adaptive and selection driven, whereas for parasitic and symbiotic unicellular species, which often experience bottlenecks, most gene-loss fixations seem to be neutral and driven by drift. The small population size of multicellular organisms, relative to bacteria, lead them to propose that genetic drift is a major driving force for gene-loss fixation. They also refer to a comparative analysis of five vertebrate and five insect species that showed a high association between rates of gene loss and that of molecular evolution for each species, thus suggesting that in general the fixation of most gene losses is driven by neutral evolution.

Various other authors are also increasingly stressing the important role of simplification in macroevolution, including cases of major evolutionary trends (reviewed by, e.g., [281]), further supporting the view that the existence of evolutionary trends does not mean that there is 'progress' in evolution. This leads us to a final point about Osborn's [287] example concerning the titanothere horns (Fig. 4.7). Osborn used such examples to defend his idea of orthogenesis because, for him, cases in which an evolutionary trend occurs in parallel in various lineages of the same clade indicate that there is an internal force driving all of the members of the clade in the same direction. Explaining the occurrence of this type of evolutionary trend in parallel across many lineages has been particularly problematic for Neo-Darwinists [159], as noted previously. However, within the context of Baldwin's organic selection, and of ONCE, such occurrences are less problematic, if one considers that the members of those lineages have often taken the same behavioral choices/shifts. For instance, there are empirical data showing that increased plasticity of ion-motive enzyme activity evolved repeatedly during multiple colonization events of freshwater by different populations of a marine copepod, likely through repeated selection on pre-existing standing genetic variation [265]. Interestingly, although his idea of organic selection could explain such examples, "Baldwin was an unabashed progressivist" in the sense that "progressivists take the perfection of the human mind and brain to be the culmination of evolution", as explained by Downes [115: 37]. As noted previously, earlier works by my colleagues and me have strongly criticized such a notion of progress, particularly the idea that there is a scala naturae leading to humans, by showing for instance that at least in terms of the number of their musculoskeletal structures, humans are actually simpler than most mammals e.g., [85, 86, 88–90, 99–105, 110]. Moreover, as explained in previous chapters, in this regard ONCE also markedly differs from Baldwin's ideas in the sense that it takes into account the common occurrence of etho-eco-morphological mismatches—including in human evolution (see Chap. 6) —which clearly contrasts with any idea of 'perfection' in nature.

Human evolutionary history does provides numerous cases of homoplasy (convergence, parallelism, and reversions). Many lineages developed similar features related to behaviors that resulted from a major behavioral shift made by the last common ancestors of those lineages several million years ago: the shift toward bipedalism (see Chap. 1). For instance, bipedalism facilitated tool use and related features such as increase in brain size: accordingly in various bipedal hominin species there were parallel trends toward such an increase in brain size and tool use. For example, absolute brain size in Neanderthals was actually greater than in *H. sapiens* on average (e.g. [371]. Therefore, the linear type of human evolution often seen in cartoons, a scala naturae from 'brute' and less bipedal to the more erect and 'wiser' species in nature, *H. sapiens*, is a gross distortion of the fossil evidence as explained by Gould [159]. Studies of evolutionary trends have been biased toward those taxa and structures in which a trend was foreseeable a priori to data analysis, and this sampling bias contributed to the idea that such evolutionary trends are more abundant than they probably are in reality (see also, e.g., [254, 314]. Therefore, on the one hand it is very important to recognize the existence of this bias as I have done in previous works [80–82, 109]. However, on the other hand it is also crucial to emphasize that a comprehensive theory of evolution must take into account and explain why there are many clear, unbiased cases of evolutionary trends extensively reported in the literature, including the one shown in Fig. 4.1 and the transitions from fins to limbs and origin of the mammalian ear (see, e.g., [4, 44, 193, 253–255, 324]; Elbe 2004; and references therein). Even if only approximately 5% of macroevolutionary changes are directional, as suggested by the large-scale, statistical paleontological survey of Hunt (2007), an integrative view of evolution needs to explain those 5%, which amount to a strikingly high number of evolutionary transitions in total, as well as the 95% of remaining ones that concern phenomena such as random walks and stasis. A major aim of the present book, and in particular of ONCE, is to help to provide rational, non-vitalistic, non-teleological, empirically-based explanations for these three as well as other major evolutionary phenomena.

Most examples of non-random, long-term evolutionary trends mentioned in this book refer to "active" or "driven" trends *sensu* McShea [254]. That is, there is a constant force operating through the various stages of the trend such as behavioral persistence. However, a second type of non-random trends was recognized by McShea—"passive diffusion" trends—in which a force that is variable across a range of morphologies might truncate the distribution of morphologies, e.g. there might be an absolute minimum body size for all members of a clade [11]. Human evolution is an illustrative example of ONCE because it has been seen as model paradigm of niche construction, and of some "active/driven" evolutionary trends.

For instance, modern humans participated in niche construction by stabilizing the local climatic environment through building houses and wearing clothing, changing their life history by developing medical treatments, and transforming their diet by the farming/cultivation of fruits and vegetables [232]. Moreover, as also noted by Lindholm and predicted by Baldwin's organic selection as well as ONCE, the secondary genetic impacts of these niche modifications—very likely mainly due to natural selection and probably combined with cases of genetic drift—are well known. The cultural habit of cooking has contributed to the accelerating en-cephalization in our lineage because our brain is far too energy-consuming to be kept by a raw diet. Cooking facilitates chewing and digestion, detoxifies, disinfects, and improves uptake of nutrients and energy, and our cranial anatomy and narrow hips are probably linked to our self-constructed dietary niche.

Moreover, the development of the human brain that was ultimately linked to our ancestors' behavioral choice of walking bipedally illustrates the importance of epigenetic phenomena affecting early ontogenetic stages under the influence of the newly built niche [263]. Humans became fully bipedal a few million years ago, and the fact that humans are born with neurobehavioral asymmetries is very likely at least partially related to the asymmetrical, epigenetic stimulation that fetal humans experience when the mother walks bipedally. The human uterus has an asymmet-rical shape that, in combination with gravity, has the effect of orienting most fetuses head down with their dorsal surface to the mother's left side [263, 299]. Consequently, differential stimulation is experienced by the left and right otolith organs of the vestibular system because of differences in shearing forces on them when the mother walks. Because this system develops early and is active during much of gestation, these sensory organs and related parts of the central nervous system that obtain input from them develop with dissimilar sensory input patterns. Therefore, differential otolith organ sensitivity, combined with the effect of vestibular stimulation on cranial and postcranial movements, can account for neonatal postural preferences. In addition, input from otolith organs projects to major neurotransmitter systems in the nervous system, which may result in other widespread asymmetrical effects on the prenatal ontogeny of the central nervous system [263, 299].

As briefly noted in Chap. 1, the importance of epigenetics in contributing to directional evolution, including that of our own species, is further emphasized by studies on South American aboriginal infants. Strikingly, in some male infants that hunt in treetops from early childhood, the angle and position of the big toe are changed to a more hand-like configuration, i.e. to a foot configuration more similar to that found in other, non-human hominoids (apes) [304]. Developmental factors, including pleiotropy (e.g. a genotypic trait affecting more than one seemingly unrelated phenotypic features) and ontogenetic integration, are also important in human evolution, and these factors further question previous adaptationist narra-tives—or "just-so stories" *sensu* Gould—about our evolutionary history. For instance, experimental studies with mice, combined with a review of the human fossil record and data on extant humans and apes, recently suggested that there was a coevolution of human hands and feet in which selection on the toes was substantially stronger than that on the fingers [312]. In other words, it is likely that changes in the foot mainly drove/led to parallel phenotypic changes in the hands, contrary to traditional "just-so-stories" postulating that hand evolution was far more important for human evolutionary history than was foot evolution.

> ***Box—Detail: "Just-So Stories"—Origin of the Term and Negative Outcomes***
> As noted by Smith [344], the term "just-so stories" refers to Kipling's *Just So Stories for Little Children* and was actually used by other authors, long before Gould made it so popular, in the same sense: to refer to un-testable ad hoc theories based on little or no evidence, in analogy to Kipling's fairytale-like creations. Unfortunately, evolutionary just-so stories are not mere anecdotes. They have profoundly influenced various fields of science, such as anthropology and psychology as well as our general culture, and thus directly or indirectly the life of many people. These include, among many others, the ideas of Freud and his followers [344], which affected numerous patients of clinical psychologists, as well as racist and eugenic ideas that have been, and continue to be, so detrimental to our species (see, e.g., [83]).

Movements to new geographical areas also provide emblematic cases of ONCE and the importance of behavioral shifts and their links to directional evolution/evolutionary trends occurring in various different lineages. For instance, regarding relatively small geographical changes in the areas in which the organisms live, fishes that move from open waters to settle in nearby cave environments tend to acquire features, over various generations and often relatively quickly in geological time, such as loss of pigmentation and eye reduction/loss (e.g., [313]). Many fishes enter caves occasionally, i.e. they normally have the behavioral plasticity to choose to enter or not, and so-called "obligate" cavefishes are therefore the progeny of those fishes that did choose to enter and settle in caves. It is the behavioral persistence of the offspring of the fishes that choose to do so that explains the strong directional evolutionary patterns through the following four main phenomena, among others. The first concerns learning, imitation, and so on, i.e. mainly social heredity *sensu* Baldwin. The second concerns epigenetic factors influenced by the external environment, i.e. by the cave environment itself, which faciliates/increases the success of life in the caves. The third has to do with natural selection, e.g. random mutations that are advantageous for life in caves will be favored, those that are disadvantageous would be selected against, and those that are neutral, such as changes in pigmentation, would no longer be selected for. It should be noted that apparently "neutral" features such as the presence of well-developed eyes—having eyes seemingly brings no direct disadvantage to the fish—are not neutral in the sense that their formation and maintenance might be energetically costly, so that energy/resources cannot be used for other tissues/functions.

The fourth phenomenon concerns direct organic selection of the parents themselves through their own behavior. That is, if the descendants do not succeed in displaying the behavior of the parents despite social heredity, e.g. if they do not learn/imitate properly, and display a different behavior such as going out of the cave, they would be very likely be left behind, a phenomenon that is well known for many animal species. Therefore, this is not a typical case of natural (external) selection where the members of a population would necessarily not survive or be able to reproduce in a certain external environment because, for instance, their morphology is not "fit" for it. It is instead due to a factor that is internal to the population itself: the behavioral choices of the parents. For example, if some young cavefish would leave the caves during the early evolutionary stages of a certain cavefish lineage, it can be easily envisaged that they could eventually find food from and/or reproduce with fishes living in the original (non-cave) habitats, if their parents would help/not neglect them. Therefore, with these four major factors, as well as others that could be combined with/added to them, such as genetic drift, it is easier to envision why so many lineages of fish have evolved similar traits, over and over again, related to life in caves.

Regarding migrations to distant geographical areas, if, for example, a certain type of seed disappears (ecological shift) in an island, or if for some reason birds arrive on a new island (very likely due to a behavioral shift by the birds to start with) where there are only larger seeds, many choices can be made by the birds. They can stay and try to eat the bigger seeds—which ultimately will lead to their death if that is not possible/is somewhat disadvantageous—or to them becoming adapted to eat bigger seeds, as seen in Darwin finches, with trial-and-error persistence combined with external selection and so on. Or, alternatively, the birds can change their niche by eating something else, or they can move away to another island where there are smaller seeds or where there are other types of seeds/foods. We know that such choices were often made in evolution: that is why animals live in many different geographical areas. The same happens with changes in environment. For instance if a certain region becomes colder, the birds can stay and for example through natural selection those mutations that will lead to feathers that would protect more against the cold would be selected, and so on. However, the birds can also choose to migrate to hotter places instead, as many birds do during their annual migrations, for instance. Therefore, once again, in all of these cases there are extremely important behavioral choices, including many behavioral shifts, and the fact that some species of birds continue to return repeatedly to exactly the same geographic region in their annual migrations shows how common behavioral persistence is and how it can lead to long-term evolutionary trends.

Another powerful example of ONCE and the links between behavioral choices/shifts and persistence and evolutionary direction/trends concerns the relationships between prey and predator or between competitors and the so-called "evolutionary arms race". I briefly wrote about this topic in Chap. 1, and will add a few comments about it here. Due to the initial behavioral choices of, for example, both a group of predators and a group of prey that lead them to co-exist in the same geographic place and time, predators can choose to hunt those specific prey among many other possible options. This often leads to an "arms race" between the two populations with the predators and prey developing traits and counter-traits that potentially allow them to hunt/escape more efficiently, respectively. In this case, this will be an example of behavioral choices that lead to behavioral persistence that, in turn, further prolongs the "arms race", and so on. Such eco-evolutionary feedbacks have been empirically demonstrated in numerous field and experimental works, occurring generally when the ecosystem effects (biotic and abiotic) of a population of organisms reciprocally influence fitness variation in that population, selection pressures, and/or evolutionary responses [246].

However, it is important to stress that such an "arms race" is not at all the only possible scenario. If their behavioral/physiological/anatomical plasticity allows them to do so, the predators can form groups to be more efficient in the hunt, such as lionesses do, or the prey can also form groups and become more difficult to catch. Or the preys can, for instance, migrate or choose to fight back, whereas the predators can shift to other preys if the original preys become too difficult to hunt. These types of behavioral choices and shifts are evident in nature because most predators do not hunt the members of a single taxon having instead a very diverse group of possible targets. It is precisely this ability to learn, the trial-and-error behaviors—as well as other more complex behaviors in organisms that can do so [75]—that allows the "arms race" to occur. For example, when the preys behave in new ways to try to escape, e.g. by running in a zigzag pattern, the predators will have to learn to adjust to that new behavior, and so on. Such cases of co-evolution are thus clear cases of organic selection *sensu* Baldwin, and it can be easily understood how they can lead to evolutionary trends, in both the predators and the prey, over long periods of time. This is seen, for instance, in cases of directional evolutionary changes such as the origin of the unique musculoskeletal features that led to the cheetahs' characteristic high-speed locomotion and fast acceleration (see, e.g., [194]).

In summary, it can be said that Baldwin's idea that organic selection can explain evolutionary trends has been increasingly supported, and recovered, by researchers from different fields in the last decades. For instance, Depew [76: 13] noted that Baldwin's idea "is that germinal and organic selection coincide to evolve a congenital instinct that is stable enough to buffer the organism's effort to respond to environmental pressures, but at the same time enough to allow further modification by ontogenetic adaptation." He notes that "what we would call a feedback loop between instincts and behaviors has been expanded to embrace a wider loop between organisms that are ontogenetically adapted and the arrow of evolution." According to Depew, this view was strongly criticized by Neo-Darwinists, in large part due to their neglect of ontogenetic processes; what authors such as Dobzhansky saw was essentially natural selection operating on adult phenotypes in populations. That is also why these Neo-Darwinists tended to put Waddington on the side of Lamarckians, because Waddington argued that genetic assimilation needs pre-existing genetic variation but defended that such variation can be triggered by formed and forming tissues in "transferring competence" during the ontogenetic process [76].

As stressed in Raff's [302: 87–88] fascinating book *The Shape of Life*, case studies from paleontology such as those concerning the Cambrian boundary show that the relatively rapid appearance of new forms/phyla is normally associated with a clear increase in the behavioral repertoire of the organisms involved in those rapid evolutionary transitions. For instance, the fossil record traces in the Vendian (a period from approximately 650 to 541 million years ago) are generally simple meanders that often cross themselves; according to the author, this is probably not a particularly efficient way for a bottom feeder to forage. By the early Cambrian (the Cambrian is a period from approximately 541 to 485 million years ago), spirals and regular meandering trails that do not cross each other emerge as do complex burrows with elaborate branches and translocations through the sediment and more complex arthropod tracks and burrows (note that the phylum Arthropoda includes invertebrates such as insects, arachnids, and crustaceans). He interprets this evidence as a coincidence of three important evolutionary transitions: (1) an increase in the diversity of animals; (2) an improvement in burrowing capacity, which very likely represents e.g. the acquisition or fine-tuning of a coelomic body structure; and (3) the occurrence of new, complex behaviors reflecting the key evolutionary innovations of more effective brains and nervous systems, which is likely related to gene-circuit and neural-circuit explosions. As Raff [302: 99] notes, in major mass extinctions, such as that at the end of the Permian (a period from approximately 299 to 252 million years ago), "something more than just replacements of unlucky groups by lucky ones seems to have occurred." That is, according to him "there has been an expansion of the ecological roles of the survivors over those of the Paleozoic fauna" (N.B.: the Paleozoic era spans from the Cambrian to the Permian, i.e. from approximately 541 to 252 million years ago).

In Raff's view, post-Paleozoic faunas are even more diverse than those of the Permian, including diverse pelagic herbivores, suspension feeders and carnivores, a wide range of epifaunal modes, and a great expansion of infaunal feeding strategies, thus reflecting e.g. the radiations of clams and sea urchins. In addition, Raff provides three other examples of such expansions in behavioral repertoires. The first is the ecological escalation seen in evolution between competing forms, especially between carnivores and their preys, which led to a long-term increase in adaptations. For instance, the number of marine families specialized for predation by shell breakage was rapidly increased in the post-Paleozoic fauna, and accordingly various mechanical defenses have in turn evolved in mollusk shells. Second, the behavioral shift related to occupation of land by animals opened the window for a wide range of novel behaviors. According to him, terrestrial organisms often face a more stressful range of environments than do marine organisms, with only seven phyla including successful land animals, all of them showing numerous anatomical and physiological changes related to life out of water. The third example is the neural revolution, i.e. sophisticated behavioral modes evolved among cephalopods (a molluscan taxon including octopuses), fishes, insects—including social insects—birds, and mammals with the brain increasing dramatically in size over the past 50 million years of the great radiation of placental mammals.

Eminent and influential paleontologists are also starting to increasingly recognize and stress that behavior and ecology play a major role in evolution in general and in macroevolutionary phenomena such as mass extinctions in particular. For instance, in his recent book Eldredge [121] linked behavior, ecological specialization, overspecialization, and extinction and explicitly recognized that ecology is the guiding principle that controls the rates of evolutionary change as proposed in ONCE. He stated: "the principle can be extended even further when we come to realize how crucial the ecological behaviors of species are in determining how rapidly their adaptations are likely to change in future evolution - and how likely species are to become extinct and to give rise to descendant species" [121: 151]. As Eldredge himself acknowledges, this statement contrasts remarkably with the line of thinking of most paleontologists for many decades, who tended to elaborate macroevolutionary ideas that were more disconnected from the organisms themselves, their behavior, and their ecology.

One interesting aspect of his book is that it argued that the more specialized a species is, the less likely it is to continue to recognize appropriate habitats as conditions change. Species displaying a less plastic/diverse inventory of behavioral choices, such as those focusing on a single type of food, are the most susceptible. For instance, parasites often have a single host species, which does not present any problems for them as long as the host species does not become extinct. That is why, according to Eldredge, ecologically specialized species become extinct at much higher rates than do ecologically generalized species in the fossil record. The concept that specialization often leads to an evolutionary dead end was first proposed by Cope [60] as the "law of the unspecialized" and has continued to be key for evolutionary biology since then [208]. In Eldredge's view, the balance of life will tend to produce ecologically specialized organisms because they often flourish more than generalists in the short run, but extinction then normally affects more the ranks of the specialists. In the long run, the generalists thus hang on—'living fossils' often being generalists—whereas the ranks of specialists are quickly refilled by the continuous evolution of new taxa. For him, taxa that descend from species that are already somewhat specialized tend to have a greater chance of focusing on a specific portion of the resources not completely exploited by the parental taxa. He designated this as a "ratchet-like mechanism" of the quick accumulation of evolutionary change as lineages keep splitting and new taxa are formed from old ones within specialized lineages. Thus, he directly connects behavioral/ecological specializations to cladogenesis and the rapid evolutionary events predicted in punctuated equilibrium. In turn, he argues that stasis is often related to generalist lineages because without a comparable degree of successful speciation, these lineages tend to have far fewer extant species at any one time than their specialized counterparts. That is, generalists are not really evolving at slower morphological rates: they are simply not generating so many new species. As a result, the gradual fluctuation of form among the members of the generalized taxa is not being fixed by cladogenesis, thus leading to new species.

Importantly, Eldredge argues that mass extinctions are probably mainly related to changes in ecological networks. That is, they are of course also often deeply related to major and random abiotic changes such as the occurrence of ice ages or the impact of meteorites. As noted previously and shown in Fig. 1.2, randomness is also considered to play a very important role in evolution, including macroevolution, in ONCE. However, Eldredge stressed that these abiotic factors are often deeply associated with major biotic changes concerning whole ecosystems, which lead to ecological collapses in which very complex ecological networks among numerous species are affected. Like developmental constraints, which can be 'positive' or 'negative'—i.e. facilitate or limit evolution (Fig. 1.2)—mass extinctions can also be seen as 'positive' or 'negative', according to him. They are 'negative' in the sense that they contribute to the limitation or termination of the evolution of taxa such as non-avian dinosaurs. But they can also be seen as 'positive' because by leading to such ecological collapses, they then facilitate the origin of different/more recent groups and ultimately of entire higher taxa with new bodyplans. As Eldredge [121: 157] put it, "it is now crystal clear that life virtually cannot evolve to any significant further degree unless extinction has come to eliminate a goodly percentage of Earth's living occupants."

Eldredge [121: 233–236] called this idea the "Sloshing Bucket" theory of evolution. That is, evolution is contingent on environmental change, and small perturbations such as fires or volcanoes may be destructive. However, in general after such events recruitment of the same/similar species from neighboring areas rebuilds ecosystems more or less similar to the ones that were destroyed. But when ecological disturbance is intense and widespread enough, entire species or even higher taxa are in danger, and regional ecosystems can collapse entirely ('turnover pulses' *sensu* Vrba). In turn, such regional extinctions can lead to episodes of speciation that result in the rebuilding of new ecosystems. Such events are, according to him, "the main locus of adaptive change in the evolution of life." Last, there are other, more severe types of events, which are the "truly global mass extinctions", resulting in greater scales of ecological disturbance and thus in greater evolutionary responses/transitions, and ultimately in larger differences in the nature and composition of the ecosystems when stability is finally recovered. As noted by him, "this picture contrasts mightily with the traditional image of the history of life progressing more or less smoothly, with new groups appearing regularly, and adaptive change accumulating in a stately and wholly regularized fashion." Therefore, I will end this chapter with a quote from Eldredge [121: 243] that emphasizes his passion for science and biology as well as the renewed focus on the organisms and their behavior, and on ecology, within the field of evolutionary biology in general and paleontology and the study of macroevolution in particular. He stated: "No scientist can safely get too far from the empirical world... I am going back to the fossil record, these days with an ecological slant, to look at the relation between stability and change that I know is typical of most species, and patterns of stability and change that may hold true for entire ecological communities... Therein should lie some solid clues as to just how the ecological and evolutionary realms are related... The beat indeed goes on."

Chapter 5
Behavioral Leads in Evolution: Exaptations, Human Evolution, Lamarck, the Cuvier-Geoffroy Debate, and Form Versus Function

For the purpose of this book, particularly this chapter, and for the form-versus-function debate, it is crucial to further discuss examples such as the emblematic one described by Osborn and shown in Fig. 4.1. As noted in Chap. 4, Osborn used this example to support his idea of orthogenesis. That is, that there is an innate—vitalistic in a way—tendency to evolve in a unilinear way due to some internal mechanism or "driving force" because random mutations could not be the drivers of such directional evolutionary trends. However, neither Osborn nor the Neo-Darwinists took into account that behavioral choices/shifts could be the main initiators and drivers of such trends as argued by authors such as Baldwin. I agree with Neo-Darwinists such as Simpson that, in theory, even very small horns could have been "better than nothing" for fighting. But that is actually an unsuitable example for the Neo-Darwinist motto of "first random mutations that lead to phenotypic variation/changes that are then selected by natural selection" (see Table 2.1). This is because fighting between the animals would have to occur first, as recognized by the phrase "better than nothing for fighting". Therefore, the first, most important event would be a behavioral choice/shift, in which a new behavior (fighting) was chosen by the organisms, and only later did random mutations that led to anatomical traits that were "better than nothing" for fighting become selected within that behavior/lifestyle, and so on. The same applies to sexual selection, which is not a type of Darwinian natural selection primarily driven by the external environment, as recognized by Darwin himself (see Chaps. 4 and 8). It is instead truly a part of organic selection as defined by Baldwin, in which a subset of a population (females) first makes a behavioral choice toward preferring mates with certain features, which subsequently drives the direction of the evolution of those features in the other sex by social heredity combined with natural selection, and so on. It is therefore striking that despite so many such obvious examples in nature and in the literature, Neo-Darwinists continue to follow the "random mutation first" motto without acknowledging that in many—likely most as I argue in this book—cases the behavior must change first.

© Springer International Publishing AG 2017
R. Diogo, *Evolution Driven by Organismal Behavior*,
DOI 10.1007/978-3-319-47581-3_5

A well-known example of how a behavioral choice of individuals of the same species led to a remarkable polyphenism concerns the evolution of horns in beetles. As reviewed by Nijhout [271: 15–16], small males of the dung beetle, *Onthophagus taurus*, are essentially hornless, whereas large males have well-developed cephalic horns. Horn length varies allometrically with body size, but this allometry is highly nonlinear, and each of the two horn morphs is associated to differences in body size. Males defend tunnels made by females and use their horns to fight other males for access to females. Males that happen to be small are at disadvantage against large males. The behavioral answer/choice of small males is therefore often to not attempt to fight but either to try to sneak past a defending male or dig their own tunnels until some distance below the defending male to thus meet the female. The alternative behavioral strategies and mating tactics of large- and small-bodied males result in a divergent selection on horn size because large males benefit from having the largest possible horns for fighting, whereas small males profit from having the shortest possible horns because horns can interfere with their sneaking tactics. According to Nijhout, horn size in this beetle species is a good example of adaptive phenotypic plasticity with two adapted extremes, the allometry of horns being a reaction norm (range of different phenotypes resulting from a same genotype: see later text) of horn size on nutrition, which is the factor linked to the size of the beetles.

These examples thus lead us to a long-standing broader debate among comparative anatomists and biologists in general: which changes first, form or function/behavior? This debate is often said to have been initiated by George Cuvier and Étienne Geoffroy Saint-Hilaire, but it actually dates back to at least the time of Aristotle and other Greek philosophers [223]. Thousands of papers and hundreds of books have been written about it, so I will not discuss its historical background and intricacies in detail here (see Gould's [159] lucid review on the subject and the many references given therein). I will instead focus on its connections to ONCE. As shown in Table 2.1, Lamarckians and supporters of Baldwin's organic selection tend to argue that function/behavior changes first, whereas Neo-Darwinists and mutationists often argue that form changes first mainly due to random mutations. This is actually an interesting, and apparently paradoxal point, because many Neo-Darwinist adaptationists in theory are said to follow the functionalist ideas of Cuvier, but Cuvier argued—*contra* the structuralist views of Geoffroy—that function determines form. That is, within Cuvier's functionalism, one would normally not expect that random changes of form would happen before functional/behavioral changes do, as is often argued by Neo-Darwinism (see Table 2.1).

> **Box—History: Aristotle, Darwin, and the Cuvier-Geoffroy Debate on Form versus Function in the Context of ONCE**
> An insightful account about the Cuvier-Geoffroy debate and its historical and philosophical roots and subsequent implications was given in Asma's [19] book *Following Form and Function—A Philosophical Archaeology of Life Science*. In short, Cuvier attributed a primary role to function/behavior,

similarly to what had been proposed by Aristotle. In this sense the fact that once of the tenets of ONCE is that there is often a 'function/behavior before form' order of events in evolution would be more in line with Cuvier's idea than with Geoffroy's one. But evidently these issues are not black and white. Firstly, for Cuvier the predominance of function/behavior was based on a teleological argument, related to final causes and God's plan, while ONCE refutes such teleological ideas and the notion of purpose, as explained in the present book. Étienne Geoffroy Saint-Hilaire was precisely trying to provide an explanation for the anatomical patterns seen in nature without recurring to a religious teleology. Geoffroy's argument that form is more relevant than function/behavior was therefore instead based on the unity of form, i.e. form is originally similar and can perform various functions, and only subsequently becomes different in different, derived taxa. As will be explained in the paragraphs below, this idea of "many-forms-to-one-function" and the related notion of exaptation are actually crucial for ONCE. So, ONCE does combines aspects of both the central ideas defended by Geoffroy and Cuvier, as it does for many other ideas or theories that have long been seen as contraditory (see Chaps. 9 and 10).

This highlights, once again, the theoretical confusions that often result from the lack of a broader multidisciplinary context—including a historical one—because this point is often confused in the literature. Neo-Darwinists essentially argue that random mutations lead to morphological variation, which is then selected by natural selection: adaptationist Neo-Darwinists specifically argue that this selection leads to adaptations to the environment/habitats where organisms live. Therefore, even for adaptationist Neo-Darwinists who supposedly follow Cuvier's functionalist ideas, variations/changes in form do normally occur before the selection is done, and thus before the functional adaptations take place. For instance, according to Neo-Darwinists, including adaptationists, there was already a variation in giraffes, with some of them having longer necks and others having shorter necks, before the ones with longer necks were selected due to the fact that they were more successful in reaching the leaves of Acacias (see Chap. 2). Therefore, paradoxically, in this respect they actually mainly agree with Geoffroy's structuralist idea in the sense that form would normally change before function/behavior, as noted above. In fact, as noted by Asma [19: 105–106], Darwin himself criticized the type of functional ideas that are often defended by adaptationist Neodarwinists. Darwin wrote: "why should the brain be enclosed in a box composed of such numerous and such extraordinarily shaped pieces of bone? As Owen has remarked, the benefit derived from yielding of the separate pieces in the act of parturition of (viviparous) mammals, will by no means explain the same construction in the skull of (oviparous) birds".

My opinion on this matter is that surely there are numerous cases in which form is modified first and can potentially lead to successful evolutionary lineages as in the case of Goldschmidt's hopeful monsters. Such cases may be particularly important in major macroevolutionary events, such as the rise of new higher taxa and/or bodyplans, as I will explain in Chaps. 8 and 9 However, I would argue that in the overall context, including both microevolution and macroevolution, successful evolutionary changes mostly involve cases in which the function/behavior was the main primary driver that secondarily lead to pronounced anatomical changes. This is the case for the beetle horns mentioned just previously, wherein the two different behavioral choices of individuals within the same species are the primary cause that led, secondarily, to an extreme morphological polyphenism. I referred earlier to the example of the brown anole lizards of the Bahamas islands, which also supports this view (Chap. 3). Gould [159] provided another emblematic example that, coincidently, involves the first taxon that I studied in detail: catfishes. Some members of the genus *Synodontis* and of other genera of the catfish family, Mochokidae, swim upside down (Fig. 5.1). This is clearly an example of 'behavior/function first' because anatomically these fishes are just like other *Synodontis*, in terms of gross musculoskeletal features, that swim normally [80]. The only major exception is pigmentation, thus reinforcing the idea that external anatomical features are in general more evolvable than internal ones and thus more often linked with the current behavior displayed and/or habitat inhabited by the organisms, i.e. less prone to eco-morphological mismatches than are for instance "deeper" skeletal and muscular features. For instance, the belly of *S. nigriventris* tends to be darker than the back. This feature can potentially be adaptive, as it makes it harder for predators to see the catfish when looking up because the predators will see the lighter colors of the catfish's back when the catfish is swimming upside-down [40, 283].

Fig. 5.1 *Synodontis nigriventris*, swimming upside down (by Gourami Watcher [*attribution* http://creativecommons.org/licenses/by-sa/3.0; https://upload.wikimedia.org/wikipedia/commons/0/0f/Featherfun_Syno4.jpg])

Another example from a group I also studied in detail, primates, concerns ca-
puchin monkeys. The musculoskeletal system of the hands of these monkeys is
completely normal for a New World monkey (i.e., monkeys that live in the
Americas), i.e. without special bones and any special muscle (i.e., same form:
contra [20]; see [88]). However, the behaviors/functions displayed by them are very
different from those seen in most other New World monkeys, especially in terms of
their extremely skillful manipulative capacities—including the common use of nut
cracking in the field—which lead humans to currently train some of them to be
'helping hands' for quadriplegics [75]. Interestingly, the opposite can be said about
gibbons and siamangs, which do not display especially skillful manipulative
capacities compared with other apes but uniquely share with humans—within
primates, due to parallelism (homoplasy)—the presence of two distinct muscles of
the forearm that attach on the thumb (extensor pollicis brevis and flexor pollicis
longus: [88]).

An endless number of other similar examples of 'function/behavior first' are
provided in the literature. For instance, Streelman and Danley [349] refer to many
microevolutionary examples supporting this idea and provide an interesting discussion on their broader implications. Regarding the fossil record, Erwin [124]
reviews numerous cases in which independent evidence indicates that behavior
changed before morphology including several cases associated specifically with
morphological novelty. As noted by Lindholm [232: 450], evidence for morphological adaptations subsequent to behavior-driven food choice is well known in the
fossil record. For example, a high-resolution record of sticklebacks from the
Miocene (epoch from approximately 23 to 5 million years ago) showed a change in
dental microwear, indicating a shift in diet from pelagic zooplankton to benthic
items, linked with a transition from pelagic to benthic habitats. This shift was in
turn associated with a corresponding enlargement of spines and body armor—
probably linked with increased predation—but preceded the increase of body amour
by 200 years. As Lindholm stated, "sticklebacks first changed diet and habitat, and
later morphology followed". Similarly, "increase in African herbivore pro-
boscideans (a clade that includes elephants) molar crown height (hypsodonty)—a
morphological adaptation to a more abrasive diet dominated by grasses—significantly lagged after the species changed habitat from woodland to open savanna
grassland during the Tertiary (Neogene) period" (the Neogene spans approximately
23–2.6 million years ago). Simoes et al. [337: 915] review a case-study from locusts
that shows that a 'function/behavior first' order of events is frequently seen not only
during evolution, but also during the development of each individual. For instance,
in locusts "behavior is the first characteristic to be modified by crowding, with
solitarious locusts acquiring key behavioral characteristics of the gregarious phase
within 4–8 h of crowding, including increased activity and locomotion and the
propensity to aggregate." The initial "behavioral shift from avoidance to attraction
sets up a positive feedback loop that facilitates all subsequent phase-related phenotypic changes and amplifies phase change from the individual to the population
level; morphological and physiological modifications only occur in subsequent

stadia or generations, with a maternal epigenetic mechanism contributing to the trans-generational accumulation of phase characteristics."

It is also striking that a 'function/behavior first' order of events should be seen as uncommon in the literature when one takes into account that most known recent evolutionary changes in our own species—which is the one for which scientists have compiled the most detailed data so far, both for the genotype and phenotype and for their links—clearly also follow this order. The reason that people living in northern regions have lighter skin is not that, first, random mutations led to such a change, completely independently of the geographic region where they lived and/or of their behavior and behavioral persistence. This is because for such mutations to be selected and thus spread with the population, one key initiator needs to have taken place previously: the behavioral choice of a group of humans to migrate—either consciously or without knowing which direction was "north" or the type of environment they/their descendants would find—which is irrelevant here as explained in Chaps. 1 and 3. These humans and their descendants persistently did migrate, and eventually, after various generations, reached geographical areas such as today's Norway or Iceland, for instance. Only then, when humans reached such northern regions and chose to stay there, through behavioral persistence did natural selection of light skin color become a particularly key player. That is, only within the context of that constructed niche—i.e. of life in those northern regions—could then the random genetic mutations that led to a lighter skin (that allowed to better cope with the very limited sun exposure and thus low ultraviolet levels and therefore avoid vitamin D deficiency) be selected and spread within the population by way of natural selection. In fact, probably similar mutations occurred randomly many times in human evolutionary history, including in Africa, but if humans would have remained exclusively in Africa, for instance, such mutations would never have been selected, and be so spread, within the members of our species.

A similar 'behavior before form' scenario also occurred in such populations that chose to migrate to and reached extremely cold regions: only after reaching those regions natural selection started to playe a crucial role by often selecting individuals with a shorter/thicker stature that minimized heat loss, thus leading to the evolutionary trend called "Allen's rule" [10]. As Youson [415: 257] stated, "as described by Bateson, each individual in a population is like a 'jukebox', having the potential to play many tunes but playing only one tune during its lifetime; the developmental tune that is played is triggered by the environment." And, I would add, also by—very likely in combination with—the behavioral choices of the organisms themselves. Another clear example concerns lactose tolerance in human adults: the genetic changes correlated to this feature were not selected and spread millions or hundreds of thousands of years ago, randomly, in the savanna or dense forests. They were instead selected, and spread, only a few thousands of years ago in pastoralist populations (e.g. [175]). That is, after—not before—the ancestors of the members of those populations behaviorally chose to herd animals.

Random mutations are random in the sense that of course they cannot be predicted, and they can occur at any time: those humans had no idea that such a mutation would arise or when it could arise. However, the fact that they are selected, and then spread in a certain population, is not random: it is normally associated with a specific niche constructed by the organisms, in which those mutations then become advantageous. Very likely, genetic changes leading to lactose tolerance have also occurred randomly many times in humans, very likely well before we began to herd animals; however, they needed a proper behavioral/ecological context to be selected and then spread in the members of our species, which only took place after humans chose to herd animals. That is why in a sense one could talk about a kind of 'constrained selection of random mutations': random in origin, successful mutations are highly constrained in the sense that they are often only selected in very specific environmental (as stressed by Neo-Darwinists) and in particular behavioral (as stressed by Baldwin) scenarios. In turn, those selected mutations allow new behaviors—in this case, they allow a substantial number of human adults to drink and tolerate milk—that will in turn potentially lead to the construction of new niches, and so on, in a complex interplay between behavioral, ecological, and genetic evolution, as postulated in ONCE (Fig. 1.2).

Many of the authors that defend a "form before function/behavior" position argue that function/behavior cannot be changed before form because a change of behavior/function necessarily requires a change in form. However, this goes against the empirical data already available showing endless cases of "function/behavior before form". I predict that numerous other cases will be revealed in the future now that increasingly more evolutionary biologists are becoming to be less biased toward/in favor of a Neo-Darwinist view of evolution. I already provided several examples in this book, such as the one about the catfishes that swim upside down (Fig. 5.1), and it is simply not possible to review all those others that I have read about, or reported in my own work, in a single book. In fact, empirical functional studies have consistently shown, over and over again, cases of "many-forms-to-one-function mapping" in a wide range of taxa, i.e. in which numerous forms can display the same function as well of cases where numerous functions can be performed by the same form. Such cases are possible precisely because of the remarkable plasticity of organisms. The combinations of behavior, function, and form are strikingly versatile and can result in an almost infinite number of possible outcomes.

Within the many empirical functional studies that could be cited here, one that I consider to be particularly powerful and illustrative, in great part due to its simplicity, was reviewed by Galis et al. [140]. In a functional/physiological study on members of the cichlid teleost fish species *Labrochromis ishmaeli*, it was shown that rapid learning can lead to fast behavioral shifts without any major change in gross anatomy. Individuals of this species were presented with thin pellets after ample experience with snails and some others with thick pellets but no experience with thin pellets or other thin food items. The first time that the fish ate thin pellets all head muscles were strongly active, as they normally are when eating thicker foods. However, by only the second time, the response of muscles such as the geniohyoideus (a hypobranchial head muscle derived from somites) was markedly weaker, as it is in other fish that are used to eating thin pellets. Based on this example and other case studies reviewed by Galis et al., they argue that flexible muscle activity patterns and reserve capacity probably supply organisms with mechanisms that allow them to undertake behavioral shifts without changes in form. This flexibility thus might in turn buffer the effects of change in form and thus reduce or avoid developmental constraints that temporarily hamper the performance of the organisms, and would also apply to other types of phenotypic plasticity. The freedom created by this plasticity and reserve capacity can thus facilitate evolutionary changes by accommodating: (1) initially, the behavioral changes; and (2) subsequently, the morphological changes that are then selected by natural selection because they are advantageous within the context of the new behaviors and/or the niches that these behaviors help to construct. Another somewhat similar study that supports this idea was published by Thompson [352], who showed that grasshoppers display substantial plasticity in head morphology in response to soft (clover) or hard (grass) food differences, which were positively linked with consumption rates (numerous other similar examples are given in, e.g., Schlichting and Pigliucci [327], and references therein; see also later text).

Another main reason that leads many authors to still not recognize the high frequency of cases of "behavior/function before form" is related to a major methodological bias that, combined with the theoretical bias in favor of Neo-Darwinism noted previously, has often led to a "self-fulfilling prophecy" scenario. The bias is that the main focus of study of paleontologists is the available morphological parts of the fossil taxa that they are studying. Because many of them are moreover biased in favor of the Neo-Darwinian idea that form often changes first, they thus tend to infer the behavior/function of those fossil taxa based on the presence of a certain form. This of course creates a circular reasoning that can easily lead to wrong conclusions. For instance, suppose that 1 million years from now a taxon of upside-down swimming catfishes acquires derived fixed phenotypes—e.g. in the spinal column—through random mutations that are advantageous for fishes displaying this peculiar behavior. If paleontologists in 2 million years knew nothing about the current behavior of upside-down swimming catfishes, they would perhaps infer that the behavior only appeared after/when the spinal column was changed, and this would not be true at all, as explained previously.

It is therefore not surprising that the already remarkably vast number of reported cases of "behavior/function before form" comes mainly from microevolutionary and experimental studies rather than from macroevolutionary/paleontological works. Of course, there are exceptions, for instance in cases where paleontologists can use other, non-morphological data to directly investigate behavior, habitat, diet, and so on, such as footprints, or trace element and stable isotope analysis. For example, the stable oxygen isotope data of the teeth of the cetacean *Indohyus*—a relative of whales—matches that of freshwater aquatic vegetation. Interestingly, apart from illustrating the fallacy of such a circular reasoning, this case also provides a further example of eco-morphological mismatch. This is because the notably long limbs of *Indohyus* had previously been interpreted—exclusively based on their form and on a priori assumptions about the links between form and behavior/function—as indicators that this taxon occupied instead a terrestrial habitat. A review of this case, and other similar examples, is given in Kemp's recent

[209] excellent book on macroevolution, *The Origin of Higher Taxa—Paleobiological, Developmental and Ecological Perspectives.*

In a recent paper, Levis and Pfennig [226] made an attempt to review key criteria and approaches that can be used to pragmatically test the hypothesis that evolution often occurs through "plasticity first". One case study discussed by them concerns cavefish, which have already been discussed in this book. According to these authors, eye loss in cave-dwelling populations of Mexican tetra (*Astyanax mexicanus*, a teleost fish) supports this hypothesis. As they state, one of the greatest abiotic differences between surface and cave environments is the lesser conductivity of cave water. When surface Mexican tetras were reared under low conductivity, they showed greater variation in eye and orbit size and upregulated the heat shock protein 90 (Hsp90). In addition, when Hsp90 was manipulated to imitate environmental stress, surface fish displayed increased variation in eye and orbit size than did the controls, contrary to cavefish, which did not amplify trait variation under Hsp90 manipulation. These and other empirical results and observations discussed by the authors suggest that stressful conditions lead to similar changes in Hsp90 function as in laboratory experiments with Hsp90 and that both stressful conditions and Hsp90 experiments result in the uncovering of cryptic variation for eye size. Moreover, individuals that develop small eyes when Hsp90 is inhibited have alleles that contribute to the inheritance of smaller eyes, even in the absence of treatment, i.e. there was seemingly a genetic assimilation. Levis and Pfennig argue that subsequent fine-tuning of this induced eyeless phenotype is linked to the "improved" functionality in cave conditions, e.g. cavefish forage "better" than surface-dwelling fish in the dark. Therefore, according to them, this case fulfils all of the criteria to be considered as a case of "plasticity-first" (in opposition to "mutation-first") evolution. Of course, they note that it is still unclear if derived (cave-dwelling) populations exhibit enhanced resource acquisition because of fine-tuning of the focal trait per se (reduced eye size) or instead due to some other phenotypic feature (e.g. olfaction) that could have arisen by way of new mutations. Be that as it may, these would anyway be new mutations selected within the niche constructed by the cavefish after their behavioral choice to live in the caves and not old, random, a priori mutations that were selected, and spread to the population, before this behavior was originated (see previous text in this chapter).

It is however important to stress again that surely there are cases of "form before function/behavior", which may be particularly important for major macroevolutionary changes such as those linked with the origin of new phyla or more "modest" cases of "hopeful monsters". However, as will be explained in Chap. 8, even in such cases behavior is very likely a key evolutionary player. One phenomenon that should, in theory, be associated with cases of "form before function/behavior" is exaptation (*sensu* Gould [159], i.e. when an ancestral form is co-opted for a new organismal behavior and/or morphological function). The famous example given by Gould concerns the proto-feathers of the ancestors of birds, which according to him were likely not related to flight per se—"how can you fly with 3% of a wing?"—but rather to thermoregulation. We now know that many dinosaurs did have wings before they supposedly acquired a "full flying" capability, so this example could seem to be a clear case of form (wing) before function (flight). However, in what may seem a paradox, in a recent paper entitled "Behavioral Leads in Evolution: Evidence From the Fossil Record", Lister [233: 326] uses this very same example to defend his idea that behavior/function usually comes before form. Lister recognizes that using the example of proto-feathers to support his idea was "somewhat stretched because the behavioral evidence for proto-flight comes not from the fossil themselves but from the behavior of morphologically-similar modern analogues." However, in an interesting twist, a deeper examination of the concept of exaptations does show that they are actually crucial to explain why "function/behavior before form" cases can be so common in evolution as postulated in ONCE. For instance, in the cichlid example of Galis et al. [140] that was cited previously, the ability of the cichlids to eat thin pellets without previously having been presented with this kind of food is related to the fact that the same form that allows them to eat thick pellets also allows them start performing a new behavior, i.e. to eat a new type of food. Therefore, this is a case of exaptation, in which behavior changed without a change of form.

> **Box—History: Exaptation—Origin of the Term and Notes on Jacob's Tinkering, Adaptationism, and ONCE**
> It is fascinating to note that the example that Vrba originally used to describe the concept of exaptation actually concerned a case in which a behavioral shift was the main driver of the evolutionary change, as recently pointed out by Eldredge [121: 49]. The example refers to the African black heron, which uses its peculiar heron-like wings to fly as all herons do. However, black herons have evolved the behavior of folding their wings into a sort of umbrella, which emits a dark, circular shadow on the water. As put by Eldredge, "fish are then attracted to this circle of darkness on an otherwise bright African day", and are then grabbed by the black herons. That is, "the wings have become a sort of trap; and, for the moment, transformed from a flying adaptation to an adaptation for feeding"—i.e. an exaptation. I completely agree with Gould in that exaptations are probably much more common than adaptationists have suggested and that exaptations further stress the enormous role played by constraints, and accordingly by tinkering (*sensu* Jacob [200]), in evolution. Therefore, exaptations do show the remarkable complexity of the interactions between form, function, and behavior that are so crucial for the ideas of ONCE.

Interestingly, recent empirical studies have questioned previous ideas—such as those defended by Gould mentioned previously—about the links between the origin of wings (the form) and of "full flight" (the function/behavior) in bird evolution. In particular, these studies contradict the idea that "proto-wings" were exclusively related to thermoregulation because they were too small to help the bird ancestors to fly. They argue that such "proto-wings" did play a major role in the first phases of the evolution of flight, which was in turn actually probably mainly driven by behavioral shifts, thus reinforcing even more the ideas of ONCE (Fig. 1.2). These studies were recently reviewed by Dial et al. [78] and Heers et al. [178] and are summarized in Fig. 5.2. These authors noted that most discussions of the origin of flight are normally embedded in (1) an adultocentric framework that assumes that adults are more affected by natural selection and that non-adult stages are mainly a means to achieve the adult form; and (2) an adaptationist framework, in which it is assumed that the peculiar forelimb traits found in adult birds are adaptions (a term that includes both adaptations and exaptations: see, e.g., [159]) for flight. According to the authors, this leads to the simplistic idea—and circular reasoning as explained previously—that animals with an apparent "flying morphology" are flight-capable and that animals without such an apparent morphology are not. Within such biased frameworks, developing (non-adult) birds were not considered to be crucial for understanding the origin of flight because, although they often have "proto-wings" that in some ways resemble those of early "winged" theropods, they supposedly lack many features that were considered to be a requirement for an advanced flight capability, i.e. they do not have an apparent "flying morphology". For instance, they do not have large wings and their musculoskeletal apparatus is more gracile and their joints less constrained [178] than are the hypertrophied pectoral muscles and robust interlocking skeleton of adults.

However, Heers et al. [178: 5] argue that empirical work on developing birds challenges both these adultocentric and adaptationist views because "fledglings with very rudimentary anatomies begin flapping and producing aerodynamic forces long before acquiring 'flight' adaptations and the 'avian' bodyplan characteristic of adults" (see Fig. 5.2). They note that "juveniles in species with a diverse array of wing-leg morphologies and life history strategies frequently recruit their legs and incipient wings cooperatively to avoid predators or reach refuges, by flap-running up slopes (wing-assisted incline running/walking) and controlling aerial descents, swimming and steaming across water, and/or jumping into brief flapping flights." Furthermore, "in at least some precocial species, the development of flight capacity (~ 18–20 days in *Alectoris chukar*, or chukar partridge) precedes the development of an adult-like musculoskeletal apparatus (close to ~ 100 days) by a substantial and biologically relevant margin; immature birds thereby call into question many longstanding views on form-function relationships in the avian body plan." Therefore, a major point against the simplistic adaptationist adultocentric frame-work is that locomotor strategies and capacities actually change dramatically throughout the development of an individual, and thus even a single organism cannot be described by a single locomotor category [78]. In fact, as stressed throughout in this book, even in a single developmental stage of a single individual, there are often many different locomotor categories and types of behavior (e.g. an adult skirrel individual jumps, runs quadrupedally on the floor, climbs trees, adopts a bipedal posture to eat nuts, and so on).

According to Heers et al. [178], behavioral shifts also played a major role in the acquisition of flight during the macroevolutionary history of dinosaurs just as they do during the acquisition of flight in the ontogeny of extant birds. Namely, the authors argue that non-adult birds bring a unique perspective to the debate on form and function because these birds show how transitional anatomies function. For instance, in behaviors such as foraging and finding refuge, when the birds are still underdeveloped and highly vulnerable, selection on locomotor performance is strong [178]. The key example that Dial et al. [78] use to address Mivart's question to Darwin—"what use is half a wing during the evolution of flight"—also concerns the development of chukar partridges, as shown in Figs. 5.2 and 5.3. When threatened, flight-incapable chicks choose to reach otherwise unreachable, elevated sites that can act as refuges by engaging their "proto-wings" and hindlimbs simultaneously to flap-run up steep surfaces (Fig. 5.3). If prevented from using their wings, chukars are not able to climb slopes that are as inclined; their wings create aerodynamic forces that enhance foot traction. Developmental changes in feather structure overlap with changes in behavioral traits and with improvements in aerodynamic performance that allow birds to flap-run up steeper and steeper slopes and eventually fly because even young birds with proto-wings produce useful aerodynamic forces.

Such cooperative use of wings and hindlimbs, and the corresponding ability to access refuges, are in fact crucial in the context of chukars' behavior because this and other precocial species leave the safety of their nest early in life and receive little parental care [78]. Another crucial point for ONCE made by Dial et al. is that the study of birds such as chukars provides further evidence that form does not need to change before function because the same form can be multifunctional, as is the case of the chukar's foot. Specifically, the hindlimb of a chukar partridge is normally capable of executing foot postures as diverse as a digitigrade posture in near-level terrain, an unguligrade posture up steep inclines, and a plantigrade posture in downhill walking (for more details about each of these postures, see [78]). Therefore, within this scenario the form is the same, and it is therefore the specific behavioral choice of the bird, at each specific time of its life, that dictates the "function" performed. However, one should note that some of the specific functional/biomechanical ideas defended by Dial, Heers and colleagues should be taken with some caution because a recent biomechanical study has put in question some of the points made by these authors [73].

Returning to Lister's [233] key paper, "Behavioral Leads in Evolution: Evidence From the Fossil Record", a particularly powerful example of how behavior has driven evolution concerns the origin of bipedalism in our species. *Australopithecus afarensis* individuals that lived approximately 3.2 MYA, like Lucy, had the necessary morphological—and very likely behavioral—plasticity to use their four limbs in a more ancestral way to move in trees as well as to undertake at least some kind of bipedalism on the ground. Clearly, Lucy was not "optimized" to walk bipedally. Nor are we, as attested by the fact that about 85% of humans still retain— as a polimorphism, mainly due to phylogenetic/developmental constraints—muscles such as the palmaris longus, which are particularly useful for grasping branches in trees with a powerful flexion of the hands, but are much less useful in bipedalism [88, 89, 96]. However, as explained in Chap. 1, by choosing to consistently walk bipedally, and persistently keeping that behavior for dozens of thousands of generations, major changes—e.g. due to random genetic mutations—that were

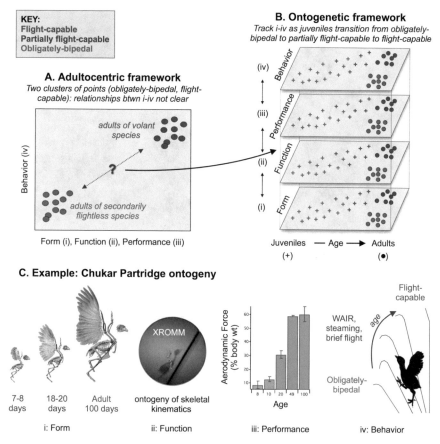

Fig. 5.2 Modified from Heers et al. [178]: According to these authors and to Dial et al. [78], from a functional perspective adult birds are the end points of an ontogenetic and evolutionary continuum and cannot clearly elucidate how specific morphological attributes affect the capacity to become airborne. This is because nearly all adult birds share a set of specialized anatomies, are flight-capable or secondarily flightless, and may not provide sufficient variation in anatomy and flight capacity to expose a relationship between these two variables. **a** In a traditional, adultocentric framework, in which many morphological features are seen as aptations (adaptations or exaptations) for aerial locomotion, most relationships between form and locomotor function are assumed rather than empirically tested. **b** Juvenile birds have rudimentary ("proto") locomotor structures and engage in pre-flight flapping behaviors as they develop into adulthood and acquire flight capacity; thus, morphing juveniles fill a gap in knowledge by helping to clarify functional attributes of the avian bodyplan, therefore revealing the form–function relationships that underlie obligately bipedal to flight-capable transitions (*i–iv*) and establishing how features are related to flight. **c** For instance, previous studies have shown that juvenile chukars with rudimentary (proto) flight apparatuses (*i*) transition from leg-to wing-based modes of locomotion by using their legs and wings cooperatively and generate small but critical amounts of aerodynamic force (*ii, iii*) that increase during ontogeny and allow birds to flap-run up steeper obstacles and eventually fly (*iv*)

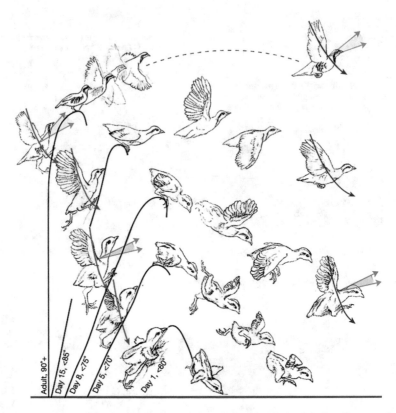

Fig. 5.3 In chukar partridges, wing-stroke is used days after hatching and during all ages and over several behaviors (i.e. flap-running, descending, and level flight): at hatch, chicks can ascend inclines as steep as 60° by crawling on all four limbs; from day 8 until adulthood birds use a consistently orientated stroke-plane angle over all substrate inclines during wing-assisted incline running (*red arcs*) as well as during descending and level flight (*blue arcs*). Modified from Dial et al. [77; for more details, see that paper]

advantageous within the context of bipedal locomotion—e.g., the big toe lying closer to the other toes, and changes in the gluteal musculature—were the subject of (external) natural selection. Random mutations that were neutral or detrimental within the context of bipedalism were very likely not the subject of such a positive selection. Thus, this continuous trial-and-error process, which was primarily driven/directed by our behavioral choices and persistence, has led to an increase in optimization between our form and our bipedal behavior, which, however, remains far from optimal, as attested by the numerous striking 'imperfections' of our body that are related to the profound constraints that come from deep time [91]. We will return to the subject of human bipedalism later in this book, but the major points to be stressed here are the following. First, our main morphological bipedal adaptions occurred *after*, and in this case were driven by, the behavior shift to bipedalism. Second, this example further shows how a ONCE view of evolution does not require any vitalistic forces or supernatural factors nor any teleological notions of "design", "purpose", "progress", or "goal" to explain long-term evolutionary trends that were consistently directional during millions of years, while dramatic environmental changes were repeatedly occurring. The first attempts of our ancestors to walk bipedally were very likely related to specific tasks of their daily lives, such as having to move from tree to tree on the ground and/or seeing their potential predators through the tall savanna grass. Those attempts were therefore surely not related to our ancestors foreseeing that, or having the purpose or final goal of, ultimately having descendents, after many generations, that would have key changes in the forelimbs and larger brains that would in turn allow them to for instance build computers or domesticate other animals. There was no long-term, "final" goal or purpose, it was simply a behavioral choice (very likely a conscious one) to walk bipedally in order to undertake much more mundane, day-to-day, tasks.

A somewhat similar example that supports the tenets of ONCE and that also concerns bipedalism comes from lizards. Clemente's [58] measurement of the body center of mass in 124 lizard species indicated that there was a significant shift among bipedal lineages and provided support for a passive bipedal model in which the amalgamation of some features led to passive lifting of the body's front. However, anatomical variation only accounted for 56% of the variation in acceleration thresholds, thus indicating that dynamics also had substantial influence on lizard bipedalism. Clemente concluded that the results show that lizard bipedalism may have first arisen as a consequence of acceleration and a shift in the body center of mass toward the back of the body, but later lineages then exploited this outcome to become bipedal more often. According to him the exploitation of lizard bipedalism was also associated with augmented rates of phenotypic diversity indicating that the behavioral choice of exploiting bipedalism may promote adaptive radiation. This example supports the ideas of ONCE in the sense that it shows, again, how sustained directional evolution—e.g. toward bipedalism—can be first driven by a behavioral shift—e.g. an increase in acceleration—which then leads to a locomotory shift—e.g. bipedalism—followed by natural selection of features that are advantageous for bipedalism within that specific niche. This in turn leads to a higher prevalence and/or more efficient type of sustained bipedalism within a clade, and so on.

Of course, behaviors such as those discussed so far in this chapter, such as the upside-down swimming of some catfishes (Fig. 5.1), are themselves plastic. Upside-down swimming is only practiced by certain members of the genus *Synodontis* and of other mochokid genera just as only some groups of hominoids (apes + humans) started to walk bipedally, and as only some human populations migrated to extremely cold regions, for instance. Because within a same taxon there are catfishes that often display upside-down swimming, whereas some others do not, the behavior is not deeply imprinted on all of the members through rigid social heredity and/or changes in their genetic makeup/genome due to genetic assimilation and/or the Baldwin effect. This catfish example thus corresponds to the very first stages of organic selection *sensu* Baldwin. That is, new members of those species of the mochokid catfish family in which the behavior is variable can either return to the ancestral behavior or adopt the upside-down swimming behavior more consistently by learning from/imitating the adult members performing the ancestral/derived behavior, respectively. In the latter case, one can easily envisage how, through behavioral persistence, the behavior can ultimately become adopted by the members of the whole population/taxon.

This discussion of plasticity leads us back to Osborn's [287] example discussed above and shown in Fig. 4.1, which in turn takes us back to the two evolutionary phenomena that are apparently paradoxical and that are explained by ONCE: the occurrence of etho-eco-morphological mismatches and that of evolutionary trends. Osborn's position on these issues was paradoxical: on one hand his theory of orthogenesis argued in favor of progress toward increasing perfection of form and function (eco-morphological match) as noted previously. In contrast, he used titanothere nasal horns as an example of how evolutionary change can cease and stasis can prevail because, for him, titanotheres may have evolved well beyond their adaptive optimum thus leading to an eco-morphological mismatch. That is, the horns might have evolved to such a degree that they interfered "with adaptation itself" [35, 169, 287]. One thing is sure: titanotheres did not reach orthogenetic "perfection" in an evolutionary context, because they all became extinct.

In such cases, extinction is therefore probably associated with loss of plasticity and with stasis, as Osborn contended, although it is difficult to imagine that the existence of the horns themselves could lead to the extinction of the whole species by negative selection. As noted by Eldredge [121: 172], one should always be skeptical about such simplistic hypotheses. For instance, for a long time it was defended that the "Irish elk" (a giant deer) evolved antlers that were so big that they contributed directly to the downfall of the species, but it is now accepted that it actually became extinct along the demise of and impressive and diverse array of other species with completely different lifestyles (see discussion on mass extinctions in Chap. 4). Be that as it may, the persistence of innovative features—such as titanotheres'horns or the so-called panda's thumb—that were advantageous within the context of, and subsequently to, organismal behavioral shifts/choices is very likely driven by behavioral persistence possibly combined with phenomena such as genetic drift. Examples of such behavioral choices/shifts would be, in this case, fighting between males, or females preferring males with big horns in titanotheres, or pandas eating bamboo. That is, the very instigators of evolutionary trends that first led to an increase in etho-eco-morphological matching—behavioral shifts/choices followed by persistence—are probably the major culprits of later etho-eco-morphological mismatches, combined with other factors (see Chap. 6).

Due to the loss of the initial behavioral plasticity that allowed the animals to adopt the new behavior in the first place, the persistence of that behavior—and thus of the epigenetic/genetic anatomical features related to it—might become counterproductive when the external environment is changed, as is the case of pandas and the current lack of available bamboo/destruction of their habitats. At one moment in time, the ancestors of pandas did not eat bamboo, but then they were able to perform a behavioral shift that ultimately lead them to eat it successfully. But now extant pandas do no longer seem to have the ability to perform a new major behavioral shift/choice and start mainly eating other foods and live in other types of habitats, as their ancestors were able to do in the past. In some cases, the needed behavioral plasticity may still be present, but the anatomical and/or genetic features that were selected within the context of the behavior acquired after the initial behavioral shift might have become too fixed to allow the body to perform potential new behaviors. For instance, the overspecialization of the "panda's thumb", which is actually formed by structures other than digit 1, might not allow pandas to revert to the more generalized type of diet that their ancestors had. That is, in this case, the "one-form-many functions" characteristic typically seen in exaptations might have been broken in the sense that the "panda's thumb" might be almost specially associated with a single function: grabbing bamboo. This emphasizes the point that behavior, function, physiology, and anatomy are all deeply interconnected. For instance, without the capacity to undertake behavioral choices, there are no behavioral shifts, and without plasticity such as that seen in cases of "one-form-many functions", organisms cannot perform certain mechanical tasks with their bodies, even if they have the behavioral capacity and drive to do so. Plasticity is thus key in the complex interplay between behavior, function, physiology, and anatomy. For instance, Smith and Redford (1990: 27) state that "anatomy is not destiny" by showing that the armadillo *Dasypoda novemcinctus* has apparent morphological specializations to eat ants and termites, but does not constrain (behaviorally) its diet to these insects, being actually broadly omnivorous, particularly in North America. As stated by Smith and Redford, such case studies highlight, once again, "the difficulties in predicting diet from morphological analysis and raise questions concerning the behavioral limits imposed by morphological specialization." I will further discuss these issues, and in particular other examples of etho-eco-morphological mismatches, in Chap. 6. It should also be noted that, apart from the phenomena discussed previously, in evolution a decrease of plasticity can also be the outcome of other phenomena, such as genetic drift (i.e. a random sampling of those organisms that had certain genetic features: see, e.g., [350]).

The ideas presented in the previous paragraphs are further supported by various types of empirical data presented in a well-documented review by Morris [265]. For instance, benthic and limnetic three-spined sticklebacks (which are teleost fishes) are relatively nonplastic regarding the anatomical features that distinguish them. However, they occupy opposite ends of the morphological reaction norm (the range of forms that organisms with a single genotype can display across different environments and developmental conditions as noted previously) of ancestral marine sticklebacks. This suggests that ancestral plasticity was necessary for this species pair to occupy their respective niches. That is, after the behavioral split (benthic versus limnetic), the plasticity became decreased within the evolution of each group. In addition, studies have shown that phenotypic change is greater in populations exposed to anthropogenic disturbance than in those that are not, mainly through plastic responses that may be adaptive. In other words, contrary to the case of pandas, some groups under stress still retain enough plasticity to change their behavior and related phenotype when that behavior puts them at risk within the context of external environmental changes, which in this specific case are related to anthropogenic disturbance.

As also noted by Morris [265], empirical data from various lineages of fish and amphibians have shown, for instance, that more plastic clades tend to be more speciose than sister taxa of similar age but with less plasticity, due to a combination of greater opportunities to diversify and augmented evolvability of plastic features and thus of decreased risk of extinction. In addition, empirical studies show that even populations that derive from ancestors that were particularly overspecialized for a certain, very specific way of life, including parasitism, have successfully changed their behavior by becoming non-parasitic and displaying a morphology that is substantially different from that of their ancestors as seen for example in lamprey evolution [415]. Morris [265] argued that random variations arising in a population may decrease plasticity and that the plastically changed phenotype linked with the behavioral shift may be negatively affected. Therefore, these variants will be likely eliminated by selection, whereas variants that decrease plasticity in the direction of the plastic change will tend to be selected and spread through the population. This may lead to a situation in which the phenotype might appear similar across generations, but its plasticity is actually increasingly reduced until an environmental shift will no longer provoke phenotypic changes. As noted by Morris, Baldwin allowed for other non-mutually exclusive scenarios to occur, such as the rise of variants that increase plasticity in general, thus increasing the 'fit' between organisms and their environment and the degree to which evolution could be directed, thus leading to evolutionary trends.

Hone and Benton [192] provided some interesting examples linking the occurrence of evolutionary trends and of developmental constraints, which can further be related to the incidence of eco-morphological mismatches and ultimately of extinctions. They specifically referred to Cope's rule, which postulates that there is a tendency for organisms to increase in size over time. This supposed tendency is probably exaggerated in the literature due to the bias of biologists, including paleontologists, to provide narratives of "trends" and "progress" (see Chap. 5). However, Cope's rule does seem to apply in at least some clades including, according to Hone and Benton, taxa in which it cannot be easily detected. This is because mass extinctions often limit the increasing of size as organisms that are too large could be for example more vulnerable to environmental crises. In fact, in some cases the evolutionary trends toward increase of size can lead to gigantism and thus to overspecialized organisms that have difficulty to cope with major environmental changes. This stresses a major point of ONCE noted previously: evolutionary trends, often driven by behavioral shifts/persistence and then secondarily directed by natural selection, can later often lead to overspecialization and loss of plasticity and thus to eco-morphological mismatches, evolutionary dead ends, and potentially to extinction.

For example, our close relative *Gigantopithecus* is an extinct Asian ape that probably lived from 9 million to 100 thousand years ago and could reach up to 3 m and 540 kg. Very likely, it could not cope with the changes during the Pleistocene epoch (spanning from approximately 2.6 million to 12 thousand years ago) from forest to savanna habitats. These changes resulted in a decrease of the main food supply of *Gigantopithecus* such as fruits, and an increase in available foods such as grass, roots, and leaves, that were dominant in the savanna and that it did not eat. According to Bocherens et al.'s [42] recent study, even when open-savanna environments were present in the landscape, *Gigantopithecus* foraging was restricted to forested habitats. Therefore, the gigantic size of *Gigantopithecus*, combined with a relatively limited dietary niche due to behavioral persistence, may have contributed to the downfall of this taxon during the dramatic forest reduction that occurred during the glacial periods in Southeast Asia. In contrast, as Hone and Benton [192] pointed out, internal constraints also probably play a major role in many cases, such as this one, in preventing an increase in size that would lead to gigantism, thus stressing another major point of ONCE: the highly constrained nature of biological evolution. One example discussed by them concerns morphological constraints on size. For example, the giraffe's neck might not be able to grow larger than it is because the giraffe would perhaps not be able to attain the needed arterial pressure for the blood to reach the head with a longer neck. In addition, developmental internal constraints limit the number of cervical vertebrae in the giraffe neck, as attested by the fact that the number of these vertebrae is almost invariably seven in all placental mammals (see previous text). That is, these constraints basically only allow—with few exceptions—the neck size of these mammals to be increased by an augmentation of the size—but not the number—of the cervical vertebrae (see Chap. 7).

In the last paragraphs of this chapter, I will focus on two papers by West-Eberhard that directly refer to many of the topics discussed so far in this book, particularly that of sexual selection. These papers connect these topics with broader evolutionary debates, including on Lamarck, and also document a fascinating—and much needed in my opinion—change in mindset among evolutionary biologists in the last decade. As pointed out by Jablonka and Lamb [198] in their book *Evolution in Four Dimensions*, Lamarck is often used as a straw-man to explain how evolution does not work. For instance, the typical example that although tails are cut in every generation of a certain dog line the tail is still present in the next generation is often provided to falsify Lamarck's idea of inheritance of acquired traits. But the idea that at least some acquired traits can be passed between generations is not wrong. A precious gem that I found on this topic, which is not often discussed in the literature, is West-Eberhard's [395] review of that book of Jablonka and Lamb [198], which not only summarizes the book but also confronts two very different ways of thinking about evolution. Of course, I highly recommend reading Jablonka and Lamb's [198] book and West-Eberhard's [393] book *Developmental Plasticity and Evolution*, both very interesting monographs, before reading West-Eberhard's 2007 review of the former book.

Although West-Eberhard's 2003 book is an excellent contribution to a less gene-centered view of evolution, particularly emphasizing the evolutionary importance of plasticity—including behavioral plasticity—it was written at a moment when most researchers were still generally reluctant to contradict Neo-Darwinism. Her 2007 review in particular reads as a passionate defense of

Neo-Darwinism, and it contrasts sharply with the very different tone she used in her 2014 publication on sexual selection, which was written at a time when several authors were no longer afraid to directly attack Neo-Darwinism. West-Eberhard [395: 440] begins by recognizing that accepting the major tenet of Lamarckism—the inheritance of acquired traits—is "acceptable if one simply remembers that there is cultural transmission of behavior patterns and agrees to call that a mode of 'inheritance'." This idea is now being accepted by increasingly more researchers—particularly the numerous ones who defend the need for an Extended Evolutionary Synthesis and who thus argue that genetic inheritance is only one of the four main types of inheritance—together with epigenetic, ecological and behavioral phenomena (Fig. 1.3). This latter idea is also defended in ONCE (Fig. 1.2), in which behavioral inheritance is seen as having a much more important evolutionary role than it does in the Extended Evolutionary Synthesis (see Chap. 1).

However, some of West-Eberhard's [395: 450] arguments explaining why she disagrees with authors such as Jablonka and Lamb, who insist "on identifying their ideas with Lamarck's" are rather odd, sometimes giving the impression that simply using Lamarck's name is a heresy. For instance, some of her arguments refer not to biological facts but instead to how one should "effectively convince" and/or "not distract" readers. She is explicit about this when she states: "part of their (Jablonka and Lamb's) defense is their interest in history, and their wish to justly acknowledge their intellectual debt to Lamarck, who, like Darwin, was neither completely correct, nor completely wrong." However, she adds, "a study of history also shows that reference to Lamarck, from example in Waddington's essays… arouses objections more effectively than it courts acceptance." She argues that referring to "Lamarckian processes may distract readers from the larger importance of environmental induction for evolution." Noting again that I profoundly admire West-Eberhard, I do think that this is an unfortunate argument. It is as though one were to argue centuries ago against citing Copernicus' or Galileo' ideas because according to those ideas the earth moves around the sun, and thus referring to them might be "less effective" and "distract" readers within the context of the accepted dogmas and the authorities that ruled at that time. The inheritance of acquired traits is not a marginal point within Lamarck's theory, just as the earth moving around the sun is not a minor point of Copernicus' and Galileo's theories: in both cases, these are central key points. And if these key points are correct, one must acknowledge that and not simply avoid talking about Lamarck just because some of his specific ideas proved to be incorrect. Most people in this planet are correct about some things and wrong about others, including Darwin as recognized by West-Eberhard and other scientists such as Einstein and Newton, so one cannot simply discard the correct ideas and theories just because their authors also had some incorrect ones. Therefore, I am optimistic that by talking about Lamarck in this chapter, and in this book in general, I will not distract you from the main tenets of ONCE, because in fact some of those tenets are related to some of his key ideas.

The second main objection of West-Eberhard [395: 444–447] was that the 'inheritance of acquired traits' might be correct, but only to a lesser extent. That is, according to her this idea was probably mainly irrelevant within the larger scheme of things in the evolution of life on this planet. She correctly points out that Jablonka and Lamb's book lacks some "clarity regarding the crucial distinction between somatic and cross-generational inheritance… for epigenetic effects… to be important for evolution they have to be passed between generations." However, in the last decade, increasingly more authors have provided data indicating that at least some epigenetic phenomena can in fact be passed between generations, i.e. that there is epigenetic inheritance apart from behavioral and ecological inheritance (Fig. 1.3) including within our own species (see, e.g., [348]). A clear, well-reported empirical example is the fact that a pregnant mother's active smoking affects not only her unborn female child but also the babies of that child even if her child and grandchildren do not smoke [199]. Numerous other examples are provided in Gilbert and Epel's [151] book *Ecological Developmental Biology—Integrating Epigenetics, Medicine and Evolution*.

An updated, fascinating list of numerous other empirical examples of non-behavioral and non-ecological epigenetic inheritance is given in Sultan's [350] recent, previously mentioned book, *Organisms & Environment—Ecological Development, Niche Construction, and Adaptation*. A particularly striking one refers to a study by Cossetti et al. [62] showing that regulatory RNA molecules released from somatic cells—namely, human tumor cells grafted into mice—intervene in the transmission of information by way of the bloodstream to germ cells and thus to sperm. If more studies confirm these results, this would therefore illustrate a way for molecular epigenetic information to become heritable as noted by Sultan. Other examples provided by her include the tracking of stable inheritance of methylation epialleles for 30 generations in a study of the plant *Arabidopsis* (a member of the mustard family) and the changes in the brain transcriptome that can be transmitted trough sperm epigenomic transmission to the third generation after rats are exposed to the fungicide vinclozolin. Importantly, these changes alter the descendants' responses to the stresses that they are perceiving and that might be completely different from those perceived by their grandparents. In fact, as pointed out by Sultan [62: 156–157], "epigenetic variation is generated at rates that are far higher than rates of genetic mutation, possibly by orders of magnitude; at the same time, epigenetic changes may often affect phenotypes more subtly than major sequence mutations, modulating gene expression in ways that may be more likely to generate useful variants."

Be that as it may, even if one would not take into account the reality of non-ecological and non-behavioral epigenetic inheritance such as those listed just above, and would merely focus on ecological and behavioral inheritance per se, the major tenet of Lamarck, i.e. that evolutionary acquired traits can be inherited, would still be correct. The question—"can acquired traits be inherited?"—is a qualitative, not quantitative, one. The answer cannot be that it is 5 or 95% correct: it is either correct or it is not. And it is correct because, as admitted by West-Eberhard and increasingly more authors, at least behavioral and ecological traits acquired in one generation can surely be passed on to the next generations (Fig. 1.3). Therefore, the most important, central tenet of Lamarckism is correct, and this fact does need to be acknowledged, as it was in Jablonka and Lamb's [198] book. Peculiarly,

West-Eberhard, a brilliant researcher who made such an effort to stress the importance of plasticity and behavior, tries to minimize, in her 2007 book review, the evolutionary importance of behavioral inheritance in order to attack Lamarck's theory. She states that "it is not clear that behavioral inheritance alone would very often lead to trait persistence across generations."

She does recognize that in theory "behavioral transmission, even if lasting only a few generations, could speed the evolution of specialization via genetic evolution when favored by selection." This is exactly the key idea defended in Baldwin's organic selection, and one of the central tenets of ONCE. But then she argues that "the primary evolutionary contribution of behavioral inheritance via maternal effects may be to bias genetic evolution, including genetic accommodation of learning as a mechanism that contributes to specialized preference." According to her "the mechanisms and morphologies of behavior are as subject to genetic variation as any other traits, and so the frequency and form of such variants will inevitably evolve as a response to selection via conventional genetic means." Therefore, she writes, "given that environmental learning is extremely widespread in animals, including in unicellular species and others that lack cross-generational transmission of learned traits, the evolutionary effect of learning for the origins of novelty is probably great, and its evolutionary effect is probably via genetic rather than via non-genetic inheritance." That is, surprisingly she mainly reduces Baldwin's organic selection to the Baldwin effect, defending the mainly gene-centered Neo-Darwinist view of evolution: anything that does not become genetically inherited (fixed) is not so relevant for evolution.

Many examples provided in the present book—and an endless number of similar ones that could have been provided, in particular from studies performed in the last decades—are demonstrating that non-genetic inheritance is in reality highly important for both microevolution and macroevolution. For instance, numerous behavioral traits that were thought to be innate—i.e. completely 'coded/fixed' genetically, such as the knowledge of newborn mammals to move toward the nipple of their mothers to obtain milk or of birds to sing—are now known to be completely dependent on epigenetic factors and/or learning or imitation as explained in Chap. 3. Furthermore, as also discussed throughout this book, learning, particularly conscious learning, is only one of the many ways in which behavioral and ecological traits are passed to the next generations as is made clear by numerous empirical examples of behavioral/ecological inheritance in taxa that are supposedly not capable of 'consciously learning' such as plants and bacteria. For example, many of the best-known empirical examples of ecological inheritance and niche construction come from plants and from social insects (e.g., [293, 327]).

Particularly interesting in this case, and the reason why I am exploring it in some detail here, is that in a paper written on sexual selection just 7 years after her 2007 paper, West-Eberhard [396: 502–507] had clearly changed her tone. This change is very likely a reflection of the increased current criticism of the very limited, gene-centered Neo-Darwinian view of evolution. West-Eberhard seems to recognize this criticism in her 2014 paper, when she states: "during the mid-twentieth century synthesis in evolutionary biology Fisher [131, 132] importantly extended

Darwin's approach to sexual selection, but there was a nearly complete blackout of the theory among organismal biologists." She noted that "this affected students like me who were steeped in the organismic evolutionary biology of the time, especially as applied to speciation and animal behavior... divergence between populations was seen as a result of selection for adaptation to ecological differences, and when it concerned sexual behavior and associated morphology the divergent phenotypes were seen primarily or even exclusively as products of selection for pre-mating reproductive isolation or species recognition." Specifically, she recognizes that, as originally contended by Darwin, sexual selection is not part of what Darwin defined as natural selection. It is instead part of what she designates as "social selection", which essentially corresponds to Baldwin's concept of organic selection (although, strikingly, she never directly refers to this latter term, nor even to Baldwin, in her paper). She states: "Darwin insisted repeatedly on the distinction between sexual selection and natural selection, with sexual selection favoring traits that benefit their bearers not from being better fitted to survive in the struggle for existence, but from having gained an advantage over other males, and from having transmitted this advantage to their male offspring alone." Therefore, by recognizing this, she contradicts one of the key ideas that she defended in her 2007 book review: that behavioral traits cannot be passed between generations for a long time. This is because there are numerous empirical examples of how behavioral preferences linked to specific, well-known cases of sexual selection have persisted for millions of years, as noted throughout the present book. In fact, she recognizes that sexual selection is often a stronger and more consistent driver of evolutionary change than is natural selection, often leading to long-term evolutionary trends that "produce complexity and exaggeration that exceeds that usually expected under natural selection."

That is, Darwin clearly distinguished between non-social and social environments, and, according to West-Eberhard, there are five main characteristics of sexual selection that effectively distinguish it from natural selection. The first refers to the "unending nature of change": "social selection promotes the evolution of extreme phenotypes in part because combat and contests of attractiveness represent unending races among rivals... the social environment for social selection changes with every improvement in a sexually competitive trait, so that there are moving targets and an unending co-evolutionary race." The second concerns the "potential for runaway change under choice": "combined with the relative lack of a ceiling for change, and the large number and diversity of stimuli that can simultaneously affect choice, the runaway process has a greater potential to have a long-term effect on sexual selection." The third is related to the "relentlessness of social selection within and across generations": "social competition within groups for access to resources—in the case of sexual selection, to mates—affects every reproducing individual of every generation in sexually reproducing species, whereas (in) natural selection... not every individual of a population may be affected by a particular kind of predator, for example, or be subjected to the same range of climatic conditions." The fourth concerns "the very large number of factors and cues that can initiate change in new directions" in sexual selection, including environmental factors, pre-existing sensory capacities of rivals and courted individuals, variations in the responsiveness of receivers, imitation, indicators of "superior" fitness, and the advantage of novelty in combat and display. Lastly, the fifth refers to the "potential for very large differences in reproductive success".

However, as originally argued by Baldwin, and also in the present book, the essential difference between sexual selection, as well as any other type of organic selection, and Darwin's natural (external) selection mainly concerns a term used in the second of her five points: the subject that makes the choice. In organic selection, the taxa that are the central subject of the evolutionary changes are the ones driving their own changes in the sense that these transformations are the result of, or at least related to, their own/their ancestor's behavioral choices in the first place (Chap. 1). This, of course, is the case in sexual selection, which usually starts with the behavioral preference of individuals of a taxon for a certain type of feature(s) in the individuals of the opposite sex of the same taxon. This contrasts with Darwin's natural selection, in which the evolutionary changes that occur in a taxon are mainly the result of external selective pressures related to abiotic factors but also to biotic factors such as whether or not an organism lives on a river including aggressive predators. For instance, the famous case of the peacock's tail made Darwin particularly "sick" because it did not fit into the context of its (external) natural selection, as noted by West-Eberhard [396: 502]. This case illustrates well the difference between what Darwin considered to be (external) natural selection versus phenomena such as sexual selection. That is, if the elaborate tail—which is mainly the product of sexual, and thus organic, selection—made the male peacocks more vulnerable to a certain (external) predator, the negative selective force of (external) natural selection would be opposed to the positive selective force of organic—in this case sexual—selection. There are many examples of such cases, and they often lead to etho-ecological and/or eco-morphological mismatches as explained previously. However, in numerous other cases, (external) natural selection secondarily reinforces the changes first driven by sexual and thus organic selection, therefore often leading to directional evolution and eventually to long-term macroevolutionary trends. These trends can, in turn, later lead to cases of mismatch due to overspecialization and/or a change in the (external) natural selective pressures, as explained previously.

A clear mark of the change in tone between West-Eberhard [396: 505] and her 2007 review of the book *Evolution in Four Dimensions* is her discussion of ideas developed within a gene-centered Neo-Darwinist context such as the "good genes" interpretation of genetic quality under natural selection made by authors such as Fisher [131]. As she points out in her 2014 paper, "extreme advocacy of the good-genes approach… e.g. … where sexual displays and morphology are described as 'sanimetric' with males usually the 'sanimeter' or 'health index' sex… takes the evolution of signals completely out of the realm of Darwinian sexual selection." She states: "although this 'eugenic' approach to sexual selection has produced exciting discoveries linking sexual display to signs of health, it has also led to amnesia regarding social selection per se." Specifically, "to call the good-survival-genes hypothesis 'sexual selection', as in 'good-genes sexual selection'… is an ironic twist seen in an historical context because Darwin so strongly emphasized that sexual selection covers phenomena not explainable in terms of natural selection". This specific example of "West-Eberhard 2007 versus 2014" shows how times are changing and how increasingly more authors are willing to frontally attack the gene-centered, Neo-Darwinian view of evolution. They are increasingly recognizing the importance of other types of inheritance, thus recovering Lamarck's idea that some acquired traits can effectively be inherited, as well as Baldwin's idea that evolutionary change is often driven by organisms and their behavioral choices, including their sexual preferences within the world of sexual organisms.

Chapter 6
Eco-morphological Mismatches, Human "Exceptionalism", Hybridization, Trade-Offs, and Non-optimality

In general, Baldwin, Neo-Darwinists, Lamarckians, and even mutationists (Chap. 2, Table 2.1) emphasized the striking "fit" (match) between the phenotype of organisms and the external environment in which they live. Of course, they all knew about the non-survival of organisms and about known cases of mass extinctions, but mostly these were associated with phenomena such as relatively rapid and/or severe changes of the external—either biotic or abiotic—environment. They did not really explore in-depth the many examples of lineages, including those that are seemingly very 'successful' in terms of taxonomic diversity, that are far from being optimal eco-morphological matches. The main reason for this oversight is that very few studies had focused on investigating and/or testing the frequency of such examples because there was a strong historical bias toward a compelling adaptationist story of how each specific feature of each taxon conferred an advantage within the habitats inhabited by the members of that taxon [159, 279]. Moreover, this general bias was often related to and/or further influenced by teleological ideas about "progress" or "purpose" in evolution, for instance toward an increase in 'perfection' of the fit between the 'design' of organisms and their environments (Chap. 1). Only more recently have authors begun to empirically and quantitatively test correlations between morphology, ecology, and phylogeny. Many, or even most, of these authors are ecomorphologists or functional morphologists who are trying to empirically find/test positive correlations between ecology and morphology that were assumed a priori by previous authors. However, and somewhat ironically, most of those studies to date have consistently revealed that morphology is actually more strongly correlated with phylogeny than with ecology, contrary to those *a priori* assumptions and in agreement with the tenets of ONCE as will be seen in this chapter.

When I refer to etho-eco-morphological mismatches, the "eco" mainly refers to "ecological" factors such as hot versus cold environments or the ecological associations between the members of a species and members of other species. That is why one can actually talk about etho-eco-morphological mismatches as an umbrella that covers mismatches between for instance the behavior and the external habitats where organisms live (etho-ecological mismatches), these habitats and the form of

R. Diogo, *Evolution Driven by Organismal Behavior*,
DOI 10.1007/978-3-319-47581-3_6

the organisms (eco-morphological mismatches), and the behavior and form (etho-morphological mismatches). Of course, because the behavior and ecology of organisms are often deeply interrelated, it is not easy to delimit what is strictly "etho" and what is strictly "eco". For instance, ecomorphologists and functional morphologist are often interested in the correlations between "function" and form so the "eco" of eco-morphology would seem to refer mostly to "function". However, historically in the form-versus-function debate, "function" was often seen instead as being more associated with behavior as is also the case in the present book (see Chap. 5).

Accordingly, the case of the pandas discussed previously is an example of etho-ecological and eco-morphological mismatches and of an etho-morphological match (morphology still accompanies persistent behavior, but both do not match the decrease in bamboo). In contrast, the example of whales—which normally still develop some hindlimb bones during their ontogeny—concerns etho-morphological and eco-morphological mismatches as well as an etho-ecological match (pelvic morphology did not change completely to optimally match the swimming behavior of whales, which does match the current aquatic habitat in which they live). Interestingly, some authors have proposed that humans, due to the behavioral shift toward bipedalism that in turn facilitated the subsequent changes leading to increasingly precise and intensive tool manufacture and use, are an example of overspecialization that might lead to our own extinction. That is, humans seem to be stuck in a persistent, monotonic behavior in which they cannot voluntarily give up/decrease their investments in technology, thus leading to a scenario in which technology, consumption patterns, overpopulation, and pollution are beginning to critically put at risk the global ecosystem without clear signals that we are able/behaviorally flexible enough to change/respond adequately to this disturbing scenario (e.g. [298]). If this proves true, it would be a major, disastrous case of etho-ecological mismatch.

Box—Detail: Mismatches, Muscles, Nerves, Medicine, Human 'imperfec-tions', and Creationism

Gould was particularly vocal about the occurrence of mismatches in living organisms because he used them to emphasize the point that organisms are not designed by a supernatural entity—as argued, for instance, by creationists —but instead are the result of a complex, constrained, contingent, and also random evolutionary history (e.g. [159]). His books provide numerous emblematic examples of such mismatches. Recently, my colleagues and I focused on examples that relate directly to the anatomy of our own species, *Homo sapiens* [97, 112], so I will refer the readers to that book and paper and not repeat here the information given there. I will just refer to a few illus-trative examples of mismatches, or "imperfections", that are found in our species. For instance, it does makes no sense, from an engineering point of view, that we have shoulder muscles that lie in our back (i.e. on the dorsal, or posterior, side of our body) but are still innervated by the ventral (or anterior) rami of the spinal nerves. It would make much more sense to have dorsal (posterior) rami innervating these muscles directly.

In fact, many of the nerve injuries we sustain in the shoulder region are related to the peculiar and often dangerous path followed by these nerves from the brachial plexus (situated ventral, or anterior, to the shoulder) to the back of the shoulder. This configuration is due to both phylogenetic and developmental constraints: we are descended from fish, in which the muscles of the pectoral appendage are mainly on the ventral part of the body, which corresponds to the so-called anterior part of the body of human anatomy. Then, during our evolutionary history, some of those muscles migrated to the dorsal part of the body that corresponds to the so-called posterior part of the body of human anatomy, and the nerves accordingly extended dorsally (posteriorly) to innervate them. One specific example is the muscle supra-coracoideus of amphibians and reptiles, which lies ventral to the pectoral girdle. This muscle gave rise to the muscles supraspinatus and infraspinatus of mammals (including humans), which are completely dorsal (posterior, in human anatomy) to this girdle and are innervated by the suprascapular nerve that has to pass through a small notch of the girdle (suprascapular notch) in order to reach the dorsal side of this girdle and innervate them [85]. The pectoral appendicular muscles, which include these and other shoulder muscles, develop ontogenetically from the ventral part of the trunk muscu-lature together with other trunk muscles, such as the intercostal muscles, that are also innervated by the ventral rami. Therefore, due to evolutionary and developmental constraints, it would in theory be very hard to maintain the functionality/development of those muscles while losing the original motor innervation by the ventral rami/derivation from the ventral trunk musculature and then gaining a *de novo* motor innervation by the dorsal rami/derivation from the dorsal trunk musculature. For more details and many other examples of "imperfections" in the human body, see [97, 112].

Apart from the examples given by Diogo et al. [112] and Diogo and Molnar [97] (see previous box), many other authors have also called attention to mismatches occurring in humans including those between some of the crucial functions of our internal organs, such as the heart, and our physiology. For instance, after providing various such examples, Noble [272: 111] wrote: "now, by contrast, we can see that life is full of design faults, false trails, and imperfect compromise; we can still wonder at the intricate beauty of life on earth, but we no longer think that its logic is the best there could be." Similarly, Lindholm's [232] recently reviewed various illustrative cases of etho-ecological mismatches in other organisms, which he designated "maladaptive behavioral syndromes" because organisms maintain sub-optimal behaviors despite significant costs. These included, among others, high activity levels despite the presence of predators as in African springboks (gazelles), newts (salamanders) opposing predatory fish, lemmings (from the clade Rodentia, which includes rats and mice) in northern alpine habitats, and exaggerated sexual cannibalism in certain species of spiders.

However, as helpful as such examples can be in stressing the importance of evolutionary and developmental constraints and our descent from other types of animals—and thus opposing views such as those defended by creationists—I will use different examples below when I refer to several empirical studies showing the existence of eco-morphological mismatches in a wide range of taxa. This is because the examples given by Gould, by us, and by the other authors mentioned in the box and the paragraphs just previous to this one mainly refer to single species. Therefore, one could surely oppose these examples with an impressive list of "one-species" cases to argue that there is a beautiful, remarkable match between the form of the organisms of a certain taxon and the habitat where they live (fish normally live in water, birds normally fly, and so on). Such cases have been historically emphasized over and over again since Aristotle, and they were in fact the ones to which humans, who often want to find "positive narratives" as noted by Gould, paid more attention. Thence came the teleological notions of "design" and "purpose" that have prevailed for a long time, and from which not even Darwin could escape due to the philosophical context and historical constraints of his epoch (Chap. 1).

Interestingly, one particular author who has continued to stress such cases of "positive" match/correlation between form and function is one of Gould's closer colleagues, Niles Eldredge, a particularly brilliant researcher who surely cannot be accused of defending a simplistic Neo-Darwinist or adaptationist view of evolution. In his recent picturesque book *Extinction and Evolution* [121: 22], Eldredge stated: "we have come, in short, to the perception that there is a design apparent in the living world; organisms are suited to their environments." However, these are abstract concepts. As my close colleague and friend Virginia Abdala, who actually defines herself as an ecomorphologist, told me one day, the million-dollar question is this: to which specific environments and/or behaviors are we referring to? During a single day, a squirrel makes behavioral choices as diverse as climbing the main trunk of a tree, delicately moving within the thin high branches of that three, jumping from tree to tree, running quadrupedally on the ground, and standing bipedally to eat nuts. Just from the window that I am facing while I write this book, I can see that squirrels also need to escape from cars when crossing the road, avoid being bitten by dogs, and so on. To which of these behaviors, and to which specific environment, is the form of the squirrel "suited"? Is the morphology of their hindlimbs particularly "adapted" to run quadrupedally, to climb trees, to stand bipedally, or to jump between trees? That is in fact a major flaw of the adaptationist framework: it is, most of all, based on a profound simplification and trivialization of life, which does not make any justice to the morphological plasticity that allows organisms to be able to perform such different tasks and behaviors and to the striking complexity and fascinating beauty of life and its evolution.

That is why in the empirical examples I will review in the remaining text of this chapter I will avoid such cases of "one-species" and "one-function" as well as more abstract examples that can be subject to "cherry-picking" and thus lead to endless discussions of "my example is better than yours". Instead, in the paragraphs below

I will refer to empirical studies that combine the four following points. First, they are a completely random, unbiased sample of a larger number of works that were found using a Google Scholar search for terms such as "anatomy", "ecology" and "phylogeny" anywhere in an article published in a journal in the last two decades and then systematically searched for all similar works on the list of references of each of those papers. Second, only works including several species, in some cases from different higher clades, were chosen. Third, among the chosen works only those that analyzed form–behavior–ecology correlations among these species/clades using quantitative tools and within a strict cladistic phylogenetic context were selected. Fourth, the subset of works finally selected were mainly written by ecomorphologists because they are not biased, *a priori*, to produce results that would support the non-adaptationist idea of ONCE: they would more likely be biased in the opposite direction.

Of course, it will not be possible to describe here the scope and results of all of the works that were finally included in that subset. Therefore, I selected studies that broadly represent the overall patterns found in those studies and that represent several major groups within a specific selected group—the vertebrates—to show how these patterns apply to both higher (more inclusive) and lower (less inclusive) clades. The group Vertebrata was chosen to be the main case study among many other animal clades that were covered by the selected studies mainly because (1) it was by far the most highly represented clade within the selected studies; (2) it is the animal clade I know the best; and (3) the results obtained for vertebrates are essentially similar to, and thus representative of, the ones obtained for the other animal clades. Still, after referring to the selected case studies from the Vertebrates, I will also provide a few examples from invertebrates and plants to show how they effectively do reveal the same general patterns, in terms of mismatches, seen in vertebrates because I do not want the readers to just take my word for it without presenting some sound empirical cases studies on non-vertebrate taxa on a subject (mismatches) that is so crucial for this book and for ONCE.

By doing this, the readers will be able to see for themselves how the empirical data obtained in those works on all of these different taxa essentially show that morphological traits that were thought to be deeply related to specific ecological/behavioral traits within an adaptationist framework have in fact almost always a stronger positive correlation with phylogenetic factors. Unfortunately, it is rather difficult to compare such empirical eco-morphological phylogenetic studies on gross morphological traits such as the length of limbs or of fleshy fruits with traits of prokaryotes (microscopic single-celled organisms, including for instance cyanobacteria and bacteria). Therefore, it is difficult to say with confidence that the same pattern is found in prokaryotes, and thus I will simply provide a short theoretical discussion about prokaryotes after dealing with the eukaryotes, in later paragraphs.

Therefore, let us now review the empirical phylogenetic studies that were selected using the methodology briefly described in the previous paragraphs. As I have done throughout the book, and as articulated in the preface, when I review these studies I will use, as often as I can, the original words of their authors. I want to make sure that

I do not alter their original meaning in order to provide a fair review and particularly to be very clear about the specific hypotheses they tested versus their results, especially because my overall take on these results is quite different from that of most evolutionary biologists, in particular adaptationists.

I will start with the paper by Vidal-Garcia et al. [368] on "The role of phylogeny and ecology in shaping morphology in 21 genera and 127 species of Australo-Papuan myobatrachid frogs." As explained by the authors [368: 182], Australian myoba-trachid frogs include two major lineages that occupy a wide range of habitats including rainforest, woods and grasslands, and extreme arid deserts. They comprise speciose genera, each with species that specialize in a broad variety of habitats, including some species that can burrow and spend extensive periods underground. According to them, these frogs thus present "an ideal group for looking at broad patterns in adaptive morphology, testing for repeated evolution of similar patterns within species-rich genera and investigating environmental correlates and phyloge-netic constraints." They inferred "the environmental niche and examined body size and shape variation displayed by all species and genera of myobatrachid frogs to test whether environmental factors determine their morphology."

Specifically, they tested two hypotheses: "(1) is the rotund, short-limbed mor-phology of burrowing frog species an adaptation to aridity and (2) are frog species from wet environments more likely to have longer legs?" Based on these hypotheses, they specifically predicted that: "(1) the species occurring in arid habitats would display more squat bodies and short limbs, (2) species from wet habitats would display stream-lined bodies with long legs and (3) species occurring in intermediate habitats would display intermediate or conservative anuran body shapes." They tested each of these predictions with anatomical and environmental datasets for all species and in a phylogenetic context to investigate whether "the different morphological patterns are constrained by phylogeny, restricting direc-tional selection." Their results showed that despite the differences among and within genera, there is no obvious climatic correlation with body size. They noted that "previous claims that rotund, short-limbed forms reduce surface area and therefore evaporative water loss in dry habitats, sound intuitively correct, but are not supported by the different geographic occurrence of certain body forms in our data" [368: 181, 188]. They summarized their results as follows: "there was no clear relationship between body size and environmental niche, and this result persisted following phylogenetic correction; for most species, there was a better match between environment/habitat and body shape, but this relationship did not persist following phylogenetic correction; our results suggest that phylogenetic legacy is important in the evolution of body size and shape in Australian anurans."

After discussing their other results, they stated that an alternative justification for the overall lack of eco-morphological correlation they found "is that evolutionary constraints have restricted directional selection; if mutational changes at genes involved in patterning the limbs or body plan have negative pleiotropic effects upon other body parts, these genes will not be selectively favored" [368: 189]. They then concluded the paper by stating that "anurans have had a highly conserved body shape pattern since at least the early Jurassic; at a broader level, there were predictable differences in body size and shape related to environmental variation in our data set; however, our data also show that the conservative anuran body size and shape that is displayed by many species is one that can work under a wide variety of environments." As explained in Chap. 5, this notion that one form allows organisms to display different behaviors, including new ones, and thus helps them construct different niches and/or live-types in different habitats, is central to ONCE. It is especially crucial for the concept that behavior/function can frequently change before form does. As also noted previously, apart from this factor and from the evolutionary/developmental constraints hypothesized by Vidal-Garcia et al., other factors can explain the occurrence of such eco-morphological mismatches/lack of predicted eco-morphological optimality. For instance, behavior persistence can lead to behaviors and thus associated morphologies that are not appropriate for the current habitats/environment in which organisms live.

For the sake of space and practicality, and also so as not to saturate the readers with so many details for each of the numerous empirical works I will review in this chapter, after providing these details for this first case study, I will now focus more briefly mainly on the results of the next empirical works that will be reviewed. For more details about each study, the readers can always—and should—refer to the original study because I provide specific references for all of them. I will thus now briefly refer to the second study, by Fabrezi et al. [129], which also refers to anurans. The authors studied the anatomy of several muscles thought to be related to anuran locomotion and lifestyle. They then tested whether there was a correlation between locomotor modes (i.e. behaviors such as hopping, jumping, swimming, and/or walking), lifestyles (i.e. life in different ecological habitats, for instance aquatic, arboreal, and/or terrestrial), and morphological traits. Their results revealed that there is not a unique combination of morphological (muscular) characters that is clearly related to any habitat (i.e. there is no eco-morphological match) or even to any locomotor mode (i.e. there is no etho-morphological match).

Still within anurans, Moen et al. [261] recently published a paper titled "Testing Convergence vs. History: Convergence Dominates Phenotypic Evolution for over 150 Million Years in Frogs". In their first test of the relative importance of convergence versus history in frog morphology, the best-fitting model was dominated by adaptive convergence. That is, each microhabitat had only one anatomical optimum independent of clade including separate optima for aquatic, arboreal, burrowing, semi-aquatic, terrestrial, and torrent-dwelling species. However, the anatomical features included in this first test were mainly external/superficial ones, such as the length of the limbs, which in theory are less subject to constraints and/or more directly influenced by the external environment as will be discussed later in the text. The only internal one was muscle mass, and more fine-grained features— such as number or attachments of muscles—were not included. However, even among those mainly external features analyzed by them, there was a substantial contribution from phylogeny. That is, although the six microhabitat-related frog ecomorphs they studied were similar around the planet, no matter where or how many times they had evolved, they did find that the species' phenotypes were generally not at the estimated phenotypic optimum for their microhabitat. They showed instead "an imprint of history" linked with a systematic bias toward the ancestral, terrestrial phenotype.

Importantly for ONCE, the results of Moen et al. [261] support the notion that behavioral/habitat shifts can often lead to evolutionary trends toward a closer match between forms for the new habitat. That is, members of younger clades living in that habitat were less optimized, whereas members of older clades living in those habitats were found to be, in general, more optimized (but never fully optimized, as emphasized by the authors). Specifically, according to the authors their results revealed that despite the widespread adaptive convergence, there is still a large "time-for-adaptation" effect on many morphological variables, i.e. species' anatomies are often offset from the "optimum" for their microhabitat, principally toward the ancestral 'optimum' (e.g. terrestriality). In their view, frog lineages have moved frequently between microhabitats during evolution, and thus many species may not have been in their current microhabitat enough time to reach their estimated morphological "optimum". For example, most microhabitat transitions are relatively recent: 77.5% are < 80 million years old, half of the total evolutionary history of frogs. Finally, the authors note that the effects of phylogeny on form may actually explain, paradoxically, the remarkable convergence that they found across frogs. For example, all burrowing anurans may be essentially similar because they still keep the basic anatomy shared by all anurans, for instance they do not resemble burrowing caecilians or lizards. That is, the limited ways to respond to selection probably lead to frequent convergence. Therefore, overall, phylogeny and constraints do seem to have played a crucial role in the evolution of the traits we see today, even among mainly external features and particularly even in a group as morphologically, taxonomically, and geographically diverse as frogs.

Continuing with amphibians, Heiss et al.'s [179] empirical study on newts (salamanders, or urodele amphibians) stresses the importance of behavioral shifts and concerns eco-morphological mismatches but at the same time challenges the way in which topics such as the occurrence of polymorphisms are seen in the literature. As noted by the authors, certain newt species change seasonally between an aquatic life and a terrestrial life as adults because they are exposed time after time to dissimilar physical conditions affecting a wide range of organismal functions. For instance, seasonally habitat-changing newts display obvious changes in skin texture and tail fin anatomy, having markedly distinct aquatic and terrestrial morphotypes and facing a major functional challenge to change between efficient aquatic and terrestrial modes of prey capture. Newts adapt quickly to such challenges because they have a high degree of behavioral flexibility: they use suction feeding in their aquatic stage and tongue prehension in their terrestrial stage. By undertaking an anatomical examination of the musculoskeletal system of the prey-capture apparatus in two multiphasic newt species—*Ichthyosaura alpestris* and *Lissotriton vulgaris*—the authors showed hypertrophy of the hyolingual musculoskeletal system in both the terrestrial morphotype of *L. vulgaris* as well as the aquatic morphotype of *I. alpestris*. That is, they showed that the seasonal behavioral and habitat shifts are accompanied by a species-dependent muscular plasticity.

Strikingly, in such a case study where the morphological changes seem to be directly related to behavioral changes and to epigenetic factors directly influenced by the external environment, eco-morphological mismatches do occur. For instance, under natural conditions suction feeding is the dominant prey-capture mode of the aquatic morphotype and tongue protraction the prevailing prey-capture mode of the terrestrial morphotype in newts. Therefore, Heiss et al. predicted that the muscles subarcualis rectus and rectus cervicis (both these muscles lie in the neck region) should become hypertrophied/atrophied in a reciprocal manner as a response to shifts in functional demands in the two morphotypes. Specifically, their eco-morphological prediction was that hypertrophy of rectus cervicis and atrophy of subarcualis rectus would be found in the aquatic morphotype and that hypertrophy of the subarcualis rectus and atrophy of the rectus cervicis would be found in the terrestrial morphotype. However, that adaptationist prediction was contradicted by their results. Both the muscle volumes and PCSAs (physiological cross-sectional areas) of the rectus cervicis and subarcualis rectus, as well as the volumes of the hyobranchial skeletal elements (which, in humans, include structures such as the hyoid bone), were significantly greater in the terrestrial than in the aquatic morphotype in *L. vulgaris*. Conversely, in *I. alpestris* the muscle volumes, PCSAs and the volumes of the hyobranchial elements were significantly greater in the aquatic than in the terrestrial morphotype. Accordingly, the changes of the hyobranchial system within morphotypes in the seasonally habitat changing newts were different as predicted.

That is, a similar pattern of quantitative morphological changes was expected in both species based on different functional demands in aquatic versus terrestrial morphotypes. However, all tested musculoskeletal hyobranchial elements hypertrophied in the terrestrial morphotype in *L. vulgaris* as well as the aquatic morphotype in *I. alpestris*, and the authors did not find any evidence for a function-based reciprocal change. They then discussed the following questions: why is there not a general pattern of hypertrophy/atrophy in both newt species, and why do newts not reciprocally hypertrophy/atrophy the rectus cervicis and subarcualis rectus, despite different functional demands between prey capture on land and in water? They recognized that they could not resolve these questions, but noted that a possible answer to the first question is that both newt species are not equally well adapted to both aquatic and terrestrial lifestyles. In other words, their results might simply be explained by an eco-morphological mismatch, i.e. in which "form" is far from optimized for at least one type of habitat/lifestyle.

I will now move on to fishes, which are the organisms that Peter Wainwright and his students and colleagues studied when they promoted the term "many forms-to-one function" that is crucial for ONCE. Wainwright et al. [388: 259–261] state that "many-to-one mapping is a ubiquitous feature of biological design… genetic epistasis produces many-to-one mapping of genotypes to phenotypes and has long been recognized as a basic property of plant and animal genetic systems… many-to-one mapping occurs between genotype and RNA secondary structure… protein structure maps redundantly onto function." They further note that "for many physiological properties of organisms there is redundant mapping of the underlying features to values of the physiological, mechanical or performance property… for example, at the level of whole-organism performance, lizards with many different combinations of hindlimb dimensions and leg muscles can have the same jumping ability." Their empirical study of labrid teleost fishes supported this idea: "in the labrid 4-bar case it would not be possible to infer jaw morphology given Maxillary KT" (output rotation in the upper jaw per degree of lower jaw rotation). They stressed that "the weak correlations between morphological and mechanical diversity in our simulations that was caused by the many-to-one mapping of 4-bar form to Maxillary KT is, by extension, discouraging for attempts to infer patterns of niche diversity from variation in morphology."

Wainwright et al. then stated that "previous authors have noted that morphology may not map closely to ecology because of the nature of behavioral or performance filters that are imposed on this relationship; our observations… on the nonlinear mapping of form to mechanics in many systems, provide an intrinsic mechanism in the relationship between form and mechanics that also can weaken this relationship." Importantly, they obtained similar results among other groups of fishes. For instance, in Collar and Wainwright's [59: 2275] study on centrarchid teleost fishes, they wrote: "we show that many-to-one mapping leads to discordance between morphological and mechanical diversity… despite close associations between morphological changes and their mechanical effects." Specifically, each of the five morphological variables on their model underlined "evolution of suction capacity… yet, the major centrarchid clades exhibit an order of magnitude range in diversity of suction mechanics in the absence of any clear difference in diversity of the morphological variables." As noted above and in Chap. 5, the occurrence of "many forms-to-one function" and "many functions-to-one form" is very important within the context of ONCE because these phenomena help to explain the common occurrence of mismatches, of exaptations, and of "function/behavior before form" evolutionary changes, which are all central tenets of ONCE.

Moving now to amniotes (a clade including reptiles and mammals), *Anolis* lizards are often seen in the literature as an emblematic example of homoplasy due to adaptations to similar habitats and are thus expected to display a high match between ecology and morphology. Strikingly, eco-morphological empirical studies revealed that even in these lizards, form and phylogeny are deeply correlated, thus supporting the importance of evolutionary constraints. This is pointed out, for instance, by Poe [296]. He explains that several authors documented striking ecological and morphological convergence in *Anolis* lizards of the Greater Antilles. For instance, various authors have defended the existence of several specific 'ecomorphs,' i.e. lizards with a supposed set of correlated ecological and morphological states associated with their particular niche. These comprise a "crown-giant" ecomorph including large species with long tails that live on the top of the trees and a "twig" ecomorph including small species with short limbs that live on narrow perches. The existence of such ecomorphs would thus suggest that characters related to, for instance, forelimb length, hindlimb length, and tail length would not be good phylogenetic markers because they would be mainly related to, and predicted by, the habitats in which the lizards live.

However, this prediction was not supported when Poe tested the phylogenetic signal of four features that are said to be ecomorph features in both the entire *Anolis* clade and only among the Greater Antillean species from which such concept of ecomorphs was initially postulated: (1) length from snout to vent (the opening through which the lizard defecates), (2) hindlimb length, (3) tail length, and (4) number of subdigital lamellae. The null hypothesis of no phylogenetic association was strongly rejected for all features in both the entire *Anolis* clade and the Greater Antillean taxa only. In a common reaction seen in almost all of the empirical studies reviewed in this chapter, the author did not hide his surprise: "this latter result is especially surprising… the presence of a strong phylogenetic correlation in the very species for which convergence has been demonstrated (i.e. within the Greater Antillean taxa only) begs for explanation" [296: 340]. He then gave possible explanations for these "surprising" results. A possible reason is "that the ecomorphs generally constitute mini-radiations within islands." A second possible explanation "is the existence of several species that defy ecomorph characterization or occupy different ecomorphs but still share derived ecomorph characters and close relationship." Another possible justification is that "some ecomorphs appear to be monophyletic across islands… in this case ecomorph characters are a better predictor of phylogeny than island locality."

Another similar empirical example, also referring to lizards, was published just a few months ago by Olberding et al. [277: 775]. In brief, they analyzed the evolution of total hindlimb length—as well as thigh, crus (leg), pes (foot), and toe length—and its correlation with habitat use and phylogeny within 46 species of phrynosomatids (a group that includes spiny and horned lizards). They took into account that sexes are usually behaviorally and morphologically dimorphic but still found that "overall, clade-level differences were more important than habitat as predictors of segment or total hindlimb length." Such empirical studies on lizards emphasize a crucial problem that has been often neglected, creating the illusion—among scientists as well as the general public—that there is almost always a very close match between behavior ("function": "etho"), form ("morpho") and the habitat where organisms live ("eco"), i.e. that there is a "design" in nature. The problem, as noted previously, is the cherry-picking of "one trait-to-one species/taxon" examples, in which for instance a certain morphological feature of a certain taxon seems to be "perfect" for a certain "habitat". This was precisely what was done for lizards such as *Anolis*, and the cherry-picked examples were then used as "evidence" of design and of "optimality" or near "optimality" in nature. However, only rigorous empirical eco-morphological phylogenetic studies, including not only that/those species/taxa but also other species/taxa of the same clade and/or living in the same geographical region, can appropriately be used to test whether specific morphological features are more correlated with ecology than they are with phylogeny, overall. In both of these two recent studies, they were not: the adaptationist predictions were contradicted by empirical evidence.

Remaining within lizards, Vitt and Pianka [370: 7877] published an interesting paper entitled "Deep History Impacts Present-Day Ecology and Biodiversity". They explained that squamate reptiles (a clade including lizards, snakes, and tuatara) are a fascinating case study for testing theories on the evolution of ecological features because "their evolutionary history dates back to the early Jurassic or late Triassic" (Jurassic spans from about 200 to 56 million years ago; Triassic spans from about 251 to 200 million years ago), they "have diversified on all major continents, and they occupy a remarkable diversity of ecological niches." They noted that one theory postulates that ecological dissimilarities result from recent factors, such as shifts in accessibility of different prey types or interspecific competition, thus predicting that niche differences arose relatively recently ("shallow-history hypothesis"). Another theory postulates that ecological differences arose early in the evolutionary history of major groups and that present-day assemblages may thus coexist mainly because of early preexisting differences ("deep-history hypothesis").

Vitt and Pianka integrated phylogenetic data with ecological information to test these theories in squamates using data on diets of 184 lizard species in 12 families from 4 continents. On the one hand, their results revealed that there were major behavioral (dietary) changes at 6 major divergence points with major macroevolutionary implications as predicted by ONCE. For instance, the most remarkable dietary divergence occurred in the late Triassic when Iguania (including iguanas and chameleons) and Scleroglossa (including geckos) split, thus leading to their occupation of very different regions of dietary niche space. This included the acquisition of chemical prey discrimination, jaw prehension, and broad foraging by scleroglossans, which allowed them to access sedentary and hidden prey that are unavailable to iguanians. That is, as stressed by the authors, "this cladogenic event may have profoundly influenced subsequent evolutionary history and diversification", thus reinforcing the idea that behavioral shifts are probably often related to speciation/cladogenesis and subsequently to major macroevolutionary divergences. In contrast, and also as predicted by ONCE, their results indicated that "such ancient events in squamate cladogenesis, rather than present-day competition, caused dietary shifts in major clades such that some lizard clades gained access to new resources, which in turn led to much of the biodiversity observed today." That is, new behaviors that become persistent, and are then further directed by natural selection, can lead to evolutionary trends and eventually to decreased plasticity. This process might finally lead to cases in which there is a much stronger correlation between form and phylogeny than between form and the current habitat where the organisms live, i.e. to eco-morphological mismatches (see previous chapters).

A different type of phylogenetic eco-morphological study of lizards was performed by Abdala et al. [1]. In this study, "form" does not refer exclusively to external and/or hard-tissue features or to more general and/or physiological traits of muscles but rather to finer morphological and morphometric muscle traits. They explain, in their introduction, that "following the tenets of ecomorphology that hold that different ecological demands lead to different organismal 'designs', it is reasonable to expect that morphological differences in muscles, tendons, and bones of the hindlimb of lizards will reflect their ecological specializations" [1: 398]. Because the arboreal environment is one of most challenging microhabitats exploited by lizards, they analyzed pedal grasping in Neotropical iguanian lizards through a morphometric study of muscles and a morphological and morphometric

study of tendons. They also examined the relative proportions of the skeletal elements involved in grasping. Specifically, they tested whether 55 hindlimb internal
features are related to habitat and/or to the phylogenetic relationships of the 23
species they studied.

Their results showed that phylogeny was the major factor associated with the
anatomy of the hindlimb skeletal characters. Interestingly, this correlation was not as
clear for most of the variables concerning hindlimb muscle and tendon morphometric
characters, thus indicating that variation of the soft tissues cannot be explained only
by phylogeny. However, neither the osteological or soft-tissue hindlimb characters
were entirely related to habitat. In fact, overall the correlations between form and
phylogeny were stronger than those between form and habitats. The authors stated,
"the prevalence of phylogeny in shaping internal morphological traits correlated with
external ones (e.g. femur and tibia length) is striking when contrasted with the
observation that, in general, changes in body size and limb and tail proportions have
been demonstrated to be associated with the evolution of locomotor performance in
different ecological settings for several clades of squamates" [1: 404].

Again, their use of the word "demonstrated" in this latter sentence should be
taken with caution because, as explained previously, such "demonstrations" are
often performed with either smaller samples or without a strict, broad phylogenetic
analysis. For instance, one of the works listed by Abdala et al. to be contrasted with
their results was that of Herrel et al. [183], which suggested that differences in the
morphology, muscle mass, and muscle-mass distribution of the limbs of *Anolis
valencienni* and *A. sagrei* were correlated with locomotor performance and style.
However, these are only two of the many species of a single genus, so it is very hard
to argue that this suggestion is really framed in a broad phylogenetic and comparative context. As noted previously, several works on *Anolis* and other lizards that
were framed in such a broad context actually showed a stronger correlation between
form and phylogeny. Abdala et al. [1] recognized this somewhat when they stated
that "the overwhelming effect of phylogeny in shaping morphology has been
repeatedly obtained in studies of various taxa" and of many different structures, such
as the external anatomy of lacertids [364], skinks [153], and liolaemids [329, 356],
the thermal biology of liolaemids [65], and the internal morphology of geckos and
liolaemids [406, 356]. Moreover, similar results were obtained for tendinous tissue
in iguanid lizards [357], and forelimb traits in tropidurid lizards [165]. Abdala et al.
[1: 397] thus conclude that "it appears that the Bauplan of the lizard pes (foot)
incorporates a morphological configuration that is sufficiently versatile to enable
exploitation of almost all of the available habitats; as unexpected as conservation of
internal gross morphology appears, it represents a means of accommodating to
environmental challenges by apparently permitting adequacy for all situations
examined." This statement is highly relevant for the idea of ONCE defended in the
present work because it further contradicts the argument that only a previous change
of form can enable a new/different behavior/function (see Chap. 5).

I will now move on to a very different group of reptiles, birds. A recent empirical
study analyzed the links between the relative proportions of wing components
(humerus, ulna, and carpometacarpus), flight style, and phylogeny in waterbirds (a

diverse group including birds such as albatrosses, diving petrels, penguins, loons, and shags, among many others) [389]. These birds exhibit substantial diversity in flight style (e.g. flapping, flapping/soaring, dynamic soaring, flapping/gliding) and foraging ecology (e.g. feeding on the wing, from the water surface, pursuit underwater). The authors examined the phylogenic signal and used ancestral trait reconstruction to test for rate shifts in forelimb proportions. Their results revealed a nonadaptationist pattern that was, once again, surprising to the authors: "different waterbird clades are clearly separated based on forelimb component proportions, which are significantly correlated with phylogeny but not with flight style" [389: 2847–2857]. Agreeing with the previous criticism about adaptationists cherry-picking "one-species-one-function" types of examples to support their views, they stated: "although changes in locomotor ecology in birds are often expected to be generally linked to changes in forelimb shape, detection of these patterns requires their consideration in a phylogenetic framework."

Moving now to mammals, I will start with marsupials, which are less studied in such eco-morphological phylogenetic studies than are placentals (the other group of extant mammals, the monotremes, is even less studied). To analyze whether the diversity of food consumed by didelphids (opossums)—e.g. fruits, small verte-brates, insects—is related to molar size and shape, Chemisquy et al. [56] used a geometric morphometric methodology to map shape onto the phylogeny of 16 didelphid genera. Then they statistically estimated the effect of diet, size, and phylogeny on molar shape. Their results were very much in line with those of other studies discussed in this chapter. Using the authors' own words, "all the analyses indicated little correlation between diet and molar shape and a strong correlation between the position of each genus on the phylogeny and molar shape" [56: 217]. They noted: "we found a strong phylogenetic effect, but the data available make the real constraints behind this historical pattern difficult to distinguish; we believe that the broad ecological niche used by most of the groups/genera studied herein (at least regarding diet) did not generate enough selective pressure on molar mor-phology to override preexisting differences that occur among clades, subsequently confounding the relationship between diet and tooth shape" [56: 232]. According to the authors, "because a large proportion of shape variation is related to phylogeny and not to size or diet, it is possible that this part of the variation is related to genetic drift, which could be translated on a macroevolutionary scale to an evolutionary model similar to a random walk or Brownian motion or other models where the change occurred at speciation."

Similar results were obtained by Magnus and Caceres [240] as explicitly stated in the title of their paper, "Phylogeny Explains Better Than Ecology or Body Size the Variation of the First Lower Molar in Didelphid Marsupials". Following the methods of Chemisquy et al. [56], but using a larger sample, they analyzed how the first lower molar evolved among didelphids by examining the two sexes indepen-dently, the influence of body size on molar shape, whether different habitats influ-ence molar shape, and the links between these factors and phylogeny. Their study included a large number of individuals (261) from 37 species representing 14 of the 19 didelphid genera, with a total of 130 female specimens from 29 species and of

131 male specimens from 36 species. Their results indicated that the shape variation of the didelphid molars is more strongly correlated to phylogeny than to body size or habitat. In their own words, "body size assumes a secondary role when influencing molar shape adaptation in didelphids, and habitat is apparently not meaningful in such a role" [240: 10]. However, they noted that although form was more strongly correlated with phylogeny, there were detectable minor differences in molar shape variation due to habitat and body size. When considered together with phylogeny, habitat seemed to be somewhat important in the adaptive radiation of didelphid lineages. For instance, this radiation is accompanied by a general shift of habitat from more ancestral and arboreal to more derived and terrestrial/scansorial environments. Accordingly, molar shape configuration evolved from a relatively small to a large trigonid, an evolutionary trend modulated by phylogeny, according to them.

Importantly for the tenets of ONCE, Magnus and Cáceres [240: 11] link these trends to mass extinction events followed by environmental changes and behavioral changes in response to those shifts. Specifically, they argue that "following the probable Miocene mass extinction, the reduction of the forest environments allowed the development of open habitats giving more chances for diversification of the surviving Miocene terrestrial marsupials via a rapid cladogenesis (or adaptive radiation)" (the Miocene spans from about 23 to 5 million years ago). For them, "this also had other evolutionary consequences by inducing a major shift in feeding habits of the family (from frugivory related to a larger talonid, to insectivory related to a larger trigonid), with lineages becoming more insectivorous with time." This change in feeding habits may be related to the behavioral tendency toward terrestriality in didelphid evolution that occurred independently in various taxa, e.g. *Monodelphis* (short-tailed-opossum) in Marmosini and *Thylamys* (fat-tailed mouse opossums) and *Cryptonanus* (gracile opoissums) in Thylamyini. The authors also point out that the strong association between molar shape variation and phylogeny does not necessarily mean that the shape of the first molar is a conservative feature within didelphids. That is, molar shape configuration changed substantially in didelphid evolution but kept a strong phylogenetic signal with the current environment, thus playing a minor role in its macroevolution but an intrinsic role in the adaptive radiation of certain lineages.

Therefore, there is again a link between phylogenetic constraints and behavioral shifts in response to environmental changes: form is not conservative within the whole didelphid clade. However, after a certain behavioral shift is completed and a subclade follows a certain evolutionary path, then its ancestors often "get stuck" in that path by a combination of behavioral persistence (e.g. being terrestrial) and natural selection (Fig. 1.2). This leads to the maintenance of a new form or the occurrence of a certain specific evolutionary trend from that form within the subclade. This can happen even if the habitat is changed and certain other behavioral traits are changed along with it (e.g. eating food item B instead of food item A), thus leading to an overall scenario in which there is a stronger correlation between form and phylogeny than between form and habitats and even between form and certain behaviors, including dietary preferences.

In a study that also included marsupials, Narita and Kuratani [267] examined the distribution pattern of mammalian vertebral formulae as a case study to investigate the contribution of developmental constraints in vertebrate evolution. Their results indicated that the changes in the vertebral formulae in eutherian mammals (which include placentals) seem to be lineage-specific; for instance most Carnivora species have 20 instead of 19 thoracolumbar vertebrae. Such lineage-specific vertebral formulae are different from the estimated distribution pattern that they calculated on the assumption of evolution exclusively related to selective pressures. Therefore, they concluded that "developmental constraints played an important role in the evolution of mammalian vertebral formulae" [267: 91]. They further noted that in some cases the lineage-specific ontogenetic constraints seemingly are not easy to surmount because members of a clade radiate but still keep the same basic bodyplan shared by all taxa in the clade. The patterns of vertebral formulae seen in marsupials, in particular, indicate a strong influence of developmental constraints. In these mammals—which are often seen as an emblematic illustration of convergence or parallelism because they have radiated into many forms and habitats in a way somewhat similar to that seen in placentals—there is a constant vertebral formulae. The authors therefore state: "if the evolution of vertebral formulae tends to change under taxon-specific selection toward the taxon-specific habitat and morphology, the constant vertebral formulae in marsupials and variable vertebral formulae in placentals could not be reconciled [267: 104]. Therefore, they conclude, "it is again much more appropriate to consider that the stasis has occurred as a result of marsupial specific (and mammalian primitive) developmental constraints that could not be overcome in the radiation of marsupial lineages."

Moving on to placentals, we find one of the very few eco-morphological studies based on an explicit broad phylogenetic analysis in which the correlation between morphology and phylogeny is not stronger than that between morphology and ecological and/or behavioral traits. However, even in this study, by Fabre et al. [128], the latter correlations are not stronger than the one between form and phylogeny either. That is, the evolution of forelimb shape in the placentals studied by the authors, the musteloid carnivorans (a group including, e.g., red pandas, otters, raccoons, and skunks), was strongly influenced by other anatomical features such as body mass as well as by both locomotion and phylogeny (Fig. 6.1). As noted by them, musteloids form a very diverse group: they display a broad variation in body mass and live in an array of habitats that are more varied than that of any other carnivoran clade; therefore these mammals are exceptionally interesting for eco-morphological studies. The authors explain that phylogeny—one of the three major factors explaining forelimb shape is musteloids, according to their results, as explained previously—was also found to be a key factor in their previous study [127] of musleloid-shape data.

For Fabre et al. [128: 603], such a strong signal "is not too surprising as, for example, all the aquatic species included in the study belong to a single clade (Lutrinae)" (which includes otters); "thus, shared ancestry, body mass and locomotor habitat are all important factors in explaining variation in limb shape across species." This strong phylogenetic signal is clearly seen in Fig. 6.1. Another key

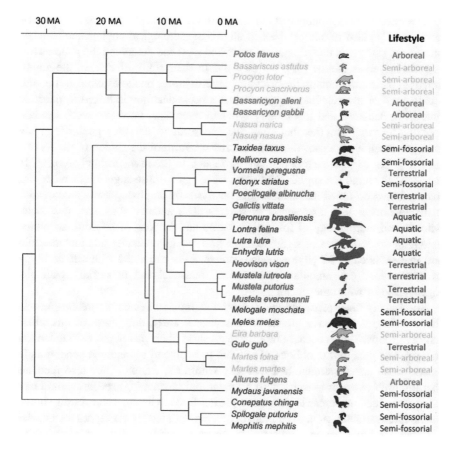

Fig. 6.1 Phylogenetic relationships of the musteloid species used in the study by Fabre et al. [128] (modified from [128]). Time scale is in millions of years; outlines used as symbols scaled relatively to body size and colors include *green* for arboreal, *yellow* for semi-arboreal, *red* for terrestrial, *brown* for semi-fossorial, and *blue* for aquatic species

point stressed by them is that this study also showed that each bone has its own ecological/functional signal, for instance "the signal provided by the shape of the humerus tends to separate arboreal and semi-arboreal species, whereas the analyses on the shape of the ulna and radius tend to separate the aquatic from the semi-fossorial species." This result reinforces the criticism made previously that the adaptationist framework is often too simplistic, not only because each organism might display several behaviors/functions (e.g. squirrels climb trees, run on the ground, stand bipedally, and so on), but also because in macroevolution there is often mosaic morphological evolution. Different parts of the body change in different ways and might be adapted for different behaviors/functions and/or ecological habitats (e.g. our hindlimbs have many features related to bipedalism, our forelimbs with tool use, our head with facial communication, and so on).

The same team that published the study on lizard muscles and tendons described previously [1] also recently published an eco-morphological phylogenetic analysis on rodents (the clade including, e.g., mice and rats) and compared the patterns seen in rodents versus lizards [52]. As they did in their study of lizards, the authors noted, in their introduction, that previous investigations on the external and internal architecture of muscles had suggested that muscle anatomy is linked to particular functional demands and locomotory types in mammals. For instance, it has been said that fossorial taxa (i.e. that often live underground, e.g. through digging) have enlarged forelimb extensor muscles and that the form of the tendon of the forearm muscle flexor carpi ulnaris is particularly related to life in a specific habitat in, for instance, didelphid marsupials. However, once again, those suggestions were often mainly based on functional morphology analyses or on phylogenetic analyses that included only a few taxa and/or characters. The authors' aim was thus to test whether the morphology of forelimb muscles and tendons of 97 adult specimens belonging to 26 species of sigmodontines (the rodent subfamily that includes mice and rats)—coded as 32 phylogenetic characters—is correlated with different types of locomotion (e.g. ambulatory, fossorial, saltatorial and natatorial), with phylogeny, and/or with other factors.

Specifically, they predicted that fossorial sigmodontines have forelimb muscles that are shorter, and have greater cross-sectional areas, than those of other locomotory groups. This is because muscles with short fibers that attach to long tendons in a pennate pattern (i.e. obliquely) are said to evolve to act as force generators for elastic-strain energy storage and recovery within the tendon. They also predicted that forelimb extensor muscles used by natatorial (swimming) species should have large cross-sectional areas to generate a greater force to thrust. Their results showed that tendon variables seem to be more correlated with locomotory types than muscle variables, but once again phylogeny is the best overall predictor of morphology. Specifically, they stated, "twelve tendon variables of the forelimb exhibit distinct differences between fossorial and scansorial sigmodontines" but "no particular morphological variables are associated with ambulatory, saltatorial, and natatorial taxa... this phylogenetic inertia could be responsible for the homogeneity in the overall muscle forelimb morphology in this group" [52: 843, 849].

We will now consider the organisms that are more closely related to us, primates. As explained in the beginning of this chapter, my colleagues and me recently compiled a series of examples showing how many of the structures of our body do not make sense unless they are seen as highly influenced by phylogenetic and developmental constraints. Many of the examples that Gould (e.g. 2002) famously used to stress the importance of constraints and to criticize both the adaptationist framework and the "argument for design" are in fact related to features that can be observed in humans, such as the so-called "blind spot" in our eyes (see [97] for more details). Therefore, what about other primates? A recent phylogenetic study about cranial morphology in lemurs by Baab et al. [21] addressed this question. In that study, phylogeny once again explained the greatest amount of variation: much smaller amounts of variation were explained by diet and even smaller amounts with activity pattern. As is usually the case, the authors noted that previous researchers had suggested that there is a positive relationship between diet, biomechanical function, and cranial shape in some vertebrate groups, but correctly pointed out: "however, some of these studies failed to explicitly incorporate phylogeny into their analyses" [21: 1444]. Furthermore, they noted that some explicit phylogenetic studies, e.g. of primates such as Old and New World monkeys, actually "did fail to recover or recovered only a weak relationship between ecological variation, including differences in diet, and cranial form (e.g. [51, 204, 250, 290])." In addition, in some studies dietary variables could predict a great proportion of variation of cranial shape, but only when phylogeny was ignored, because when it was taken into account many fewer anatomical features remained associated to ecology (e.g. [291]). Based on the results of their lemur phylogenetic study, Baab et al. [21: 1472] thus stated that "overall, lemur cranial morphology retains a strong phylogenetic signal… the correlation between diet and cranial form is weak when the underlying phylogeny is taken into account, a pattern also documented in other vertebrate lineages." According to the authors, "this may mean that diet has not strongly impacted cranial form, but may also be the result of an evolutionary history characterized by a relatively small number of dietary (behavioral) shifts that occurred in conjunction with the divergence of major clades and few instances of dietary convergences between these clades."

   In another recent study that also comprised strepsirrhines (the sister group of all other extant primates that includes lemurs and lorisoids such as lorises and galagos as well as rodents and marsupials) Ruth et al. [317] tested the usually accepted assumption that the foramen magnum position is an indicator of bipedalism because it is supposed to favor a more "balanced" position of the skull. They examined the connection between the angle of this foramen and locomotion in these three clades including bipedal or orthograde species and quadrupedal or pronograde species. Their results revealed that in marsupials and strepsirrhines there is no association between this angle and locomotor pattern. They did find a significant difference in this angle between quadrupedal and bipedal rodents, but when these taxa were analyzed in the context of enlarged auditory bullae, this relationship was no longer significant. Using their own words, they concluded that "taken together, these data indicate that several developmental modules of the cranium influence FMA (foramen magnum angle), but that locomotion does not; we caution that basicranial evolution is a complex phenomenon that must be explored in the context of each taxon's unique evolutionary and developmental history" [317: 45].

   My colleagues and I recently studied the evolution of head and limb muscles in primates. Among the non-facial muscles that make up the majority of the muscles in the head, the same pattern holds: within mammalian clades, such as primates, phylogeny is a much better predictor than ecology or behavior of the number of muscles and muscle bundles [88, 89]. But what about the evolution of the facial muscles, i.e. those muscles innervated by the facial nerve (cranial nerve VII) that are directly related to facial expression? Faces represent a more external anatomical aspect of the mammalian head that, in some groups such as primates, and particularly in humans, is deeply related to social communication. Primates rely on facial-coloration patterns for species and individual recognition and on facial expressions to evaluate the behavioral intent of conspecifics within social groups ([98, 323]). Whereas facial expressions are produced by the facial musculature, facial-coloration patterns are underlain by the pigmentation of the skin and fur. Due to the close physical proximity of these components and their mutual importance in social communication, it would be reasonable to envisage that the evolution of such facial features would be mainly related to behavioral and ecological features. However, our empirical studies have shown that although both variation in facial color, pattern, and expressions are partially explained by evolution to sociality and habitat, even supposedly highly adaptive traits composing the primate facial anatomy and function—such as facial color patterns, facial musculature, and repertoire of facial expressions—carry a strong phylogenetic signal (e.g. [98, 323]).

The number of facial muscles is, in particular, strongly influenced by phylogeny, and evolutionary changes in this number seem to have occurred at a very slow pace when compared with coloration and mobility ([86, 88, 89, 98, 101, 102, 323]). For instance, the total number of muscles/muscle bundles is much more similar in primates from a same clade (e.g. strepsirrhines, New World monkeys, Old World monkeys, or hominoids) than it is in primates from similar ecological groups. In fact, recent studies in five species of the lesser ape family Hylobatidae (gibbons and siamangs) showed that even superficial external facial-expression properties—such as number of facial expressions per unit of time and/or different types of facial expressions used—are more related to phylogenetic inertia than to socio-ecological factors [326]. As expected, however, non-facial head muscles are in general much less variable with respect to muscle presence/absence in primates than are facial muscles, which are one of the few muscle groups in which humans do have a greater number of muscles than most other primates have [86, 88, 89]. This fact seems to underline the importance played by facial communication (expression) in primate and human evolution.

In other words, for each muscle group of the head (e.g. facial, masticatory, infrahyoid, suprahyoid, and laryngeal), the number of muscles is mainly explained by phylogeny, but the differences between the number of muscles present in different primate clades are much greater concerning the facial muscles than regarding other head muscles. This disparity illustrates an important fact that, in my opinion, is not often recognized in the literature: more superficial structures, such as the facial muscles, may effectively be more versatile than more deep structures, such as the pharyngeal muscles, but this does not necessarily mean that changes in the superficial structures are necessarily "better" explained by ecology per se and more related to specific adaptations. The variability of facial structures can, for instance, be profoundly linked to ontogenetic constraints (e.g. facial bones are mainly derived from neural crest cells, whereas neurocranial bones are derived from the paraxial mesoderm). Moreover, facial muscles follow a developmental pattern that is markedly different from that followed by most other cranial muscles. They are second-arch migratory muscles that, unlike other head muscles, often insert onto bones that derive from neural crest cells of other arches (e.g. onto the mandible, which derives from the first-arch) and even onto non-skeletal structures such as the skin (see, e.g., [85]).

To show that many of the results found in the works on the Vertebrata that were mentioned in the previous paragraphs also apply more broadly, I will now refer to one of the numerous empirical phylogenetic eco-morphological studies on invertebrates that reveal patterns essentially similar to those usually found in the other invertebrate studies that I reviewed. The study, by Law et al. [221], is specifically focused on the polychaete family Opheliidae (a family of small annelid worms). The authors explain that there are considerable anatomical and behavioral differences between two major groups within this family: for instance, some animals burrow by peristalsis (through wave-like muscle contractions), whereas others display undulatory burrowing. The authors therefore analyzed the anatomical differences that might be related to these distinct burrowing behaviors and undertook a DNA-based phylogenetic analysis to test the connection between behavior, form, and the habitats where these worms live.

The mapping of the anatomical features onto the phylogeny revealed close links between morphology (namely the musculature) and behavior, and a lower correlation between form and habitat/ecology. They state that "even though the mechanical responses of muds and sands to burrowers are substantially different—muds are elastic materials through which most worms extend burrows by fracture, whereas sands are non-cohesive granular materials, suggesting that morphologies and behaviors of burrowing animals might be distinct between these two habitats - our data showed that habitat distribution is variable and did not coincide well with burrowing mode, musculature, or presence of septa" [221: 557]. They further note that "the nearly identical morphologies, musculature, and undulatory burrowing behavior within Ophelininae did not coincide with a single sediment distribution; even generalizations based on similar morphologies and musculature that appear to be convergent seem to be an unreliable indicator of habitat distribution." Various phylogenetic studies of several clades of herbivorous insects also point out that insect host-use evolution is often quite conservative regarding the host taxon, being often better predicted by phylogeny than by apparent ecological opportunity [208].

Last, I will refer to two examples from plants that summarize the pattern commonly found among the numerous plant eco-morphological studies found in the literature as documented, for instance, in the exceptional book *Phenotypic Evolution—A Reaction Norm Perspective* by Schlichting and Pigliucci [327]. As stated by these authors (pp. 182–183), there are many examples "in the literature (of plants) of cases in which patterns usually interpreted as the results of natural selection are in fact reinterpreted to be due to some sort of constraint; again, in these cases phylogenetic analyses are pivotal." For instance, they review a large study of 910 angiosperms that tested the assumption that features of fleshy fruits—such as length, diameter, mass, or energy and protein content—evolved in response to the animals that disperse those fruits. The study showed that within the 16 traits examined, 61% of the overall variance was explained by phylogeny and that most correlations between fruit traits and type of disperser were nonsignificant after considering phylogenetic effects. As they note, "the overall conclusion is that historical contingencies are generally responsible for the observed macroevolutionary scenario, while selection played a comparative minor role." They state, however, that "it is worth cautioning against over-interpretation of such analyses based on phylogenetic correlations… phylogenetic constraints are in fact a combination of ecological (or selection) and genetic (or 'historical') forces."

I understand the argument that Schlichting and Pigliucci [327] are trying to make in this latter statement. I also realize that readers can use the text given on their pages 183–188 as a counter-argument against the idea defended in the majority of the works mentioned in this chapter, that phylogeny is often a better predictor of morphology—particularly of more internal traits—than is ecology. However, I cannot completely agree with their two points about this issue, which then lead them to their oversimplified example about "selection" versus "phylogeny". In my opinion, eco-morphological mismatches should not be seen as a contest between "historical contingencies" versus "selection", nor should ecological forces simply be equated with selection. There are many other factors involved in the occurrence of eco-morphological mismatches, including a crucial one that the authors—and, unfortunately, most current researchers who also promote an "Extended Evolutionary Synthesis" (Fig. 1.3)—tend to neglect: behavioral persistence. As noted in the first chapters of this book, it is striking that such an "extended" synthesis, despite including concepts such as niche construction, does not consider behavioral choices and persistence in discussions about major evolutionary patterns. As additionally stressed in those chapters and throughout this book, within the more complex interplay of factors assumed in ONCE (Fig. 1.2), natural selection itself can be a major player in the occurrence of phylogenetic inertia and thus of eco-morphological mismatches. Therefore, any dichotomy between "phylogeny" versus "selection" does not stand within the framework of ONCE.

Despite disagreeing with the authors on this specific point, I think that Schlichting and Pigliucci's [327] book is an exceptional one. I learned a lot from it, and it reviews an impressive number of empirical examples that are extremely important within the context of ONCE. Its lack of emphasis on behavior and on the importance of behavioral choices and persistence in evolution is mostly a result of the historical context in which the book was written and of the current mindset within Evo-Devo. Among the fascinating empirical studies they reviewed, a series of studies on features that were assumed to be closely related to pollination success in *Dalechampia* (a genus of plants including, e.g., the winged beauty shrub) provide further examples of mismatches in plants and have important theoretical implications for ONCE. In those studies, various characteristics related to inflorescence were examined, and the results showed that the likelihood that the pollinators (bees of different sizes) would come in contact with the stigma (the tip of a carpel/several fused carpels of a flower) was closely associated with the distance between the stigma and the resin gland (the source of food for the bees). That is, the largest bees can contact all stigmata but focused their choices on species with large glands, which are positively correlated with the gland–stigma distance. Importantly, as noted by Schlichting and Pigliucci [327: 279], "the adaptive surface predicted from these data, however, indicated that for optimal pollen receipt, the gland-stigma distance should be short and the resin gland should be large… however… the positions of populations are often far below the predicted optimal values." They use this example to stress the importance of simultaneous selection on multiple traits, a point that is crucial to understand eco-morphological mismatches and that goes, once again, against the simplistic adaptationist assumption that each morphological feature was selected for a particular "purpose"/function.

This point in turn leads us to the concept of so-called "evolutionary trade-offs", which were also the subject of an interesting discussion in Schlichting and Pigliucci's [327] book and which bring us back again to a key issue of ONCE: generalist versus specialist strategies (see Chap. 4). As noted in page 272 of their book, numerous works have provided extensive discussions on this subject; the authors provide references for some of these works, to which readers should refer for more details. In particular, they note that some authors published empirical evidence to support the idea that generalists (phenotypically plastic individuals) will be favored over specialists ("canalized individuals") in fine-grained environments, whereas specialists are at an advantage in homogeneous environments. They also point out that various models of adaptation that are not framed on an extreme adaptationist framework do presuppose that there are costs to plasticity in cases of adaptation to variable environments. For instance, adaptation to a second habitat can involve a loss of fitness in the first habitat and, conversely, there should also be a "trade-off" in performance in other habitats by specialists. However, under ONCE not all cases of eco-morphological mismatches are necessarily related to trade-offs. In fact, even in those cases in which the two phenomena are related, I much prefer to call it a "mismatch" than a "trade-off" because the latter term reinforces both a teleological and an adaptationist—and thus, in my judgment, an incorrect—view of evolution, as explained in the box below.

Box—Definitions: 'Trade-off' Versus Mismatch, Teleology, Adaptationism, and Human Evolution

In my opinion, the term "trade-off", when used in an evolutionary context, reinforces a teleological view of evolution as illustrated by the definition of "trade-off" in the Oxford Dictionary: "a balance achieved between two desirable but incompatible features." There is nothing "desirable" in evolution: there is no final goal or aim. Using "trade-off" thus also reinforces, paradoxically, a chief idea of an extreme adaptationist framework: the concept that life is a constant "struggle". This is because the term "trade-off" gives the impression that there is always a trend toward—or a "desire" to reach—an "optimal" (desirable) fitness between the phenotype and the current habitat or habitats where organisms live, and that this is the only possible way for organisms to survive and reproduce. The idea of constant struggle is in fact supported by Schlichting and Pigliucci [327: 68] who, although not following an extreme form of adaptationism, defend the idea that "from an organism's standpoint any sub-optimal environment represents some kind of 'stress'."

However, if you have been in Africa and seen wild animals there, you have probably seen lions sleeping relaxed for huge amounts of time and small lion cubs joyfully playing among themselves. Adaptationists can argue that the sleeping lion is stressfully sleeping because he must replenish energy to catch the next prey or that the cubs are in reality playing in a struggle to learn/train how to kill prey as soon as possible. However—at least for me, perhaps a too naïve non-adaptationist researcher—this does not match my observation of the lions and their cubs. At least at that moment, for the lion and the cubs, there is seemingly no stress, no struggle. Nor does it cause stress for me—on the contrary—when I am, for instance, swimming in the sea during my summer holidays, despite the fact that I am using a body that is surely not "optimally adapted/designed" to move in the water.

In fact, as noted previously in this chapter, the fact that in humans the nerves that innervate the back shoulder muscles take a peculiar—and dangerous in terms of risk of injury—path is not due to a "trade-off" in the sense that our body is "adapted" to the "desirable" different habitats we occupy, ways of life we have, or functions we perform. It is simply because of developmental and phylogenetic constraints that are often linked to contingency, as explained previously [97, 112]. Having such a configuration provides nothing adaptive for any of the environments we live/things we do, nor has it ever did in human evolution: since humans diverged from chimpanzees, approximately 6 million years ago, it has always been a mismatch. It is an example of "bad" design—or "imperfection", related to constraints and tinkering—that was acquired hundreds of million years ago, during tetrapod evolution, and that persists today. Even the original pattern of innervation of the pectoral and pelvic appendicular muscles by the ventral rami, acquired in early vertebrate evolution, was very likely not specifically related to an "adaptation" to the specific habitats in which our ancestors lived at that time. It was instead likely related to developmental constraints, i.e. to the simple fact that the paired appendicular muscles are mainly derived from the embryonic primordia that also give rise to ventral trunk muscles such as the intercostal muscles (i.e. from "hypomeres") and that are thus accordingly innervated by the ventral rami. It is therefore, instead, yet another example showing how important a role constraints play in evolution.

In summary, the empirical examples reviewed in this chapter show that phylogenetic relationships are often a better predictor of morphology—particularly of more "internal" anatomical traits such as details about the specific attachments of muscles, number of musculoskeletal structures, and so on—than are behavioral and/or ecological traits. Because one of the main fields of my research is biological anthropology, I will further illustrate this point by referring to an subject that was briefly discussed in previous chapters and to which readers can easily relate, because it concerns human evolution: the behavioral choice made by our ancestors a few million years ago to walk bipedally. In a very simplified way, it can be said that when the environment in East Africa became drier, our ancestors could no longer find food, such as fruits, by moving from tree to tree as they had done in the dense tropical forests and as many other primates still do in other parts of Africa and in other continents. In the new savannah environment, they had to come down from the trees and move along the ground to find other trees with fruits available. In response to that environmental change, humans could have behaved in many different ways: among all animals that lived in that area at that time, including other primates, humans were the only ones that became "fully bipedal" since then, after all. Apart from having the needed behavioral, physiological, and anatomical plasticity to be able to walk bipedally, organisms also need to make an active choice to do so. During the millions of years that have passed since that initial behavioral choice, we have occupied many different habitats, from northern icy regions near the poles to high mountains, from islands to deserts, and even regions with very dense tropical forests in which our closest relatives (the apes) continue to live. Still, due to our behavioral persistence, we continue to walk bipedally, even when we go into those dense forests to observe some of those apes and various other species of primates, despite the fact that it would probably be more "efficient"—at least for certain moments—to move in the trees in such dense forests. I can attest this myself because when I went to observe the locomotion and tool use of wild chimpanzees in Uganda's Kidale forest, it was particularly difficult to follow these apes when they were moving on trees. Our bipedal walking was completely inefficient in such dense forests, and the only way we were able to follow these apes in a not completely unsuccessful way was due to the help of the local guides who were continuously cutting through the dense vegetation with machetes. That is, bipedalism is surely not the most optimal type of locomotion that an animal could undertake for every environment we inhabit within the context all of the possible theoretical biomechanical options available.

Therefore, if one were to study the correlations between, for example, the type of locomotion, the musculoskeletal morphology, the phylogeny, and the different habitats of humans (e.g. deserts, islands, icy regions, mountains, dense forests, and so on), one would clearly find a stronger link between phylogeny and form than between form and each of these types of habitats. One would conclude that the fact that phylogeny so closely matches morphology is due to "phylogenetic inertia" because the peculiar anatomical features associated with human bipedalism are only found in the descendants of a specific ancestor group within the human lineage that originated a few million years ago. It is true that this inertia is very likely partially explained by a potential decrease in anatomical plasticity/variation and/or the fixation of derived anatomical features, possibly due to the selection of random mutations. However, the crucial point that is often neglected is that the main driver of this evolutionary history, which had to take place *before* this natural selection occurred, was actually the first behavioral shift toward bipedalism in a savannah environment millions of years ago, and—crucially—our subsequent and continuous behavioral persistence to continue to be bipeds since then. This is one of the major reasons why it is so problematic in adaptationist, eco-morphological, and functional studies to always try to relate form to the current environment/habitat where organisms live:

the negligence of the crucial evolutionary role of behavioral persistence and thus of the potential—and frequent—occurrence of eco-morphological mismatches.

Accordingly, within a systems biology framework, our species now displays—in great part also because of that behavioral shift toward bipedalism and subsequent behavioral persistence to do so—a particularly striking capacity for homeostasis (greater independence from the external environment). That is, ecomorphological correlations that perhaps made sense earlier in our evolution would not make sense at all today. For example, it is often said—and supported by some empirical evidence—that humans who live in cold environments tend to have shorter extremities and a lower surface area–to–volume ratios (e.g. Bergmann's rule and Allen's rule). However, with the current patterns of human globalization, one can easily find first- or second-generation emigrants from warm regions of Africa that still have in general a higher surface area–to–volume ratio, despite the fact that they now live in Nordic countries. Likewise, we can find people who came from cold regions, with a lower surface area–to–volume ratio, living in warmer countries. Air conditioning, closed buildings, clothes, and other inventions, together with behavioral persistence, allow humans to do so. Therefore, if one were to perform an eco-morphological study on the correlation of limb length and the average outside temperature recorded for the geographical regions where we currently live, one would find endless cases of mismatches. Everybody would agree with this example, I think. The problem is that many people would say: well, but that is an exception; it is the human exception because we behave in a special way. Some people even say that we are not evolving anymore. But we are. We are evolving in a way mainly driven by our behavioral choices and persistence, and constrained by developmental and phylogenetic factors and by a complex interplay of other factors, including randomness, as are all other living beings. "Human exceptionalism"—the belief that some features only apply to humans—was actually one of the major reasons for the resistance against ideas such as Baldwin's organic selection, as explained in previous chapters.

Related to this topic, I noted previously that it is difficult to directly compare the eco-morphological phylogenetic studies reviewed in this chapter about plants and animals with works done in the unicellular prokaryotes (including, e.g., bacteria). However, what would we predict about these latter organisms within the context of ONCE and/or a systems biology framework? According to the major tenets of ONCE (Fig. 1.2), one might predict that eco-morphological mismatches are also commonly seen in prokaryotes and that form is often strongly related to phylogeny in these living beings. However, if one takes into account that eukaryotes such as animals and plants are generally thought to have stronger homeostasis than most prokaryotes, it is possible that eco-morphological mismatches are less marked/frequent in prokaryotes. As noted throughout this book, numerous factors, separately or in combination, can lead to eco-morphological mismatches, so it is possible that other factors would lead to a higher frequency of such mismatches in prokaryotes, contrary to predictions stemming from a systems biology and/or standard Neo-Darwinist point of view. Clearly, direct comparisons between phylogenetic eco-morphological studies of eukaryotes and prokaryotes are badly needed. I plan to address this particular issue in more detail in future collaborations with researchers working mainly on prokaryote organisms.

Before ending this chapter, I would like to briefly refer to a subject that is largely neglected in macroevolutionary discussions but that provides interesting insights about the links between behavioral choices/persistence, developmental constraints, phylogenetic inertia, randomness, and the occurrence of eco-morphological mismatches: hybridization. The empirical example I will mention concerns the low (about 7%) versus high (about 62%) frequency of black coats in wolves of the North American icy tundra versus the forested areas of the Canadian Arctic, respectively [14]. The differences between the wolves living in these two habitats are not yet completely clear, and some studies give conflicting suggestions about the specific order of events that many have led to these dissimilarities (see, e.g., [63]). However, genetic studies generally agree that a key gene responsible for color, beta-defensin, is lacking some nucleotides in animals that form a black coat and that the occurrence of this mutation in North American black wolves is likely a heritage from hybridization with dogs (e.g. [14]). Therefore, at least three different behavioral choices seem to be deeply related to this case study: (1) the choice of humans to migrate across the Bering land bridge; (2) their choice to take dogs with them and/or the choice of the dogs to accompany them; and (3) the choice of dogs/wolves to copulate with each other or to force members of the other taxon to copulate (behaviors similar to human raping are seen in various non-human organisms: see, e.g., [288]). Only then did North American wolves acquire this specific phenotypic variation (grey *vs.* black coats) that was subsequently selected differently in different environments: a seemingly higher selection of black coats in forested areas than in the tundra lead to the disparate (62% *vs.* 7%) frequency of black wolves in each of these respective niches, as noted previously. As expected, according to a mainly adaptationist framework, researchers and in particular the media have been mostly interested in learning why the black coats might be advantageous in forested areas. For instance, it has been said that whereas the dark coat likely does not camouflage the wolves from predators, because wolves have few natural predators, it might help them to not be seen by their prey in dark forests, in contrast to what happens in the whiter tundra. It has also been suggested that the mutated gene might increase the immune defense against infectious agents that occur chiefly in warmer forests (see, e.g., [63]).

However, I am more interested in the opposite question: why do 7% of the tundra wolves have a black coat? This could be seen as an eco-morphological mismatch because it would make the wolves more visible to their prey in the whiter tundra. From what was said just previously, the answer might well be that the tundra black wolves might have advantages in terms of a better immune defense, in a further example of pleiotropy (a genotypic trait affecting more than one seemingly unrelated phenotypic features, as explained above). That is, the answer is probably not related to the color of the coats per se but to features that are developmentally connected in less obvious ways with the occurrence of black coats. Therefore, in addition to the behavioral choices and subsequent persistence that originally led to the hybridization that produced black coats in wolves, constraining developmental factors such as pleiotropy likely contributed to the occurrence of the few cases of 'black coat in icy tundra' that are apparent eco-morphological mismatches.

New studies on hybridization and its importance for biological evolution, including the creation of new species, present problems for Neo-Darwinism and, although Darwin was clearly aware about cases of hybridization, also to his general ideas, e.g. for his vision of a mainly bifurcated "tree of life". Recent studies have stressed that the tree of life includes, in several cases, reticulation between different lineages not only among unicellular prokaryotes—which are by far the most taxonomically diverse organisms—but also among eukaryotes, including plants and animals [280]. One of my favorite examples of eukaryote reticulation is documented by the long-term study of the Grant's on the Darwin finches because it further stresses the close connection between behavioral shifts and hybridization. Grant and Grant [163] reported their observations of an immigrant medium ground finch (*Geospiza fortis*) male on Daphne Major Island that was different from the main population due to its slightly different beak morphology and peculiar song pattern. This male gave rise to a new lineage in which the females only reproduced with males of the same lineage, suggesting the start of a new species based on learned induced variation. Namely, this new lineage was established when the young immigrant male arrived on the island in 1981 and bred in 1983; this male was identified genetically as a fortis–scandens backcross to *G. fortis* that had that originated from the nearby large island of Santa Cruz and bred with a resident female of similar hybrid constitution in 1987 [164].

For the next three generations, members of this lineage bred with each other and with *G. fortis*. Only two individuals, a brother and a sister, survived the drought of 2003 and 2004, and they bred with each other when the drought finished in 2005. Their offspring then bred with each other for the next three generations and not with *G. fortis* or *G. scandens*. The hybrid lineage thus mainly functioned as a new species, being distinct in song, morphology, and genotype, and breeding endogamously, increasing in number to nine breeding pairs in 2010 [164]. Using the Grant's own words, although components of the sexual barrier among two species are often flexible, species often remain different because the behavioral imprinting mechanism on which this barrier is based is robust. Behavioral alteration of mate-signal learning by way of a peak shift mechanism, without genetic shift, may thus be expanded as an element in the evolutionary response of organisms to the challenge of survival and choosing a mate in a species-rich environment. Even among our closest relatives, the non-human primates, there is extensive hybridization, as has been well documented, for instance, in baboons by Rebecca Ackermann and colleagues. It has now been confirmed by ancient DNA studies that even in the more recent evolutionary history of our own genus *Homo*, there were cases of hybridization that had important implications for both our genotype and phenotype and for disease (e.g. Ackermann et al. 2015). I will return to the issue of hybridization when I discuss Goldschmidt's hopeful monsters in Chap. 8.

As a conclusion of this Chap. 6, I would like to emphasize a common feature among the phylogenetic eco-morphological studies reviewed in the chapter: their authors often admitted that they were profoundly surprised by their own results.

This is because they were often searching for clear, positive eco-morphological correlations that should, according to their assumptions, be stronger than phylo-morphological correlations. The authors' surprise sends us an important message: the lack of clear eco-morphological correlations is probably even much more common that an overall review of these studies indicates. This is because these authors probably represent a minority of researchers who are brave enough—and honest enough—to contradict their own predictions/*a priori* ideas, as well as to explicitly admit that they were genuinely surprised when they first noted this contradiction. It is very likely that many authors who obtained similar results—or, in particular, "worse" results, such as a complete lack of any type of correlation between morphology and ecology—simply opted to not publish their results.

Such so-called "negative results" are often ignored because there is no "story to be told", a phenomenon that is well known in science and was strongly criticized by Gould in his famous "Cordelia's dilemma" metaphor (e.g. [158, 159]; see also [84]). Therefore, based on the results of the eco-morphological empirical studies reviewed here and on the similar results of many other works that I did not have the space to review, and on this bias against the publication of such "negative" results, I think that the common occurrence of eco-morphological mismatches is probably one of the most crucial untold stories in evolutionary biology. The story remains untold because it goes against the main ideas defended by followers of Darwinism, of Lamarckism, of Neo-Darwinism, and of Baldwin—who were in one way or another influenced by the notion that organisms are "designed" to live in the habitats they inhabit—and even of the Evo-Devo Extended Evolutionary Synthesis (Fig. 1.3), as explained previously.

In my opinion, ONCE is the first attempt to specifically provide a sound, empirically-based, comprehensive, multidisciplinary explanation for the wide occurrence of such mismatches/non-optimality (Fig. 1.2). For instance, it integrates Evo-Devo notions of constraints. It also considers the crucial importance of behavioral choices, shifts, and persistence in evolution, as proposed in Baldwin's organic selection, but it takes into account a fact that Baldwin did not explore in depth: behavioral persistence can very often lead to etho-ecological, etho-morphological, and/or eco-morphological mismatches. In addition, ONCE acknowledges the importance of randomness in evolution as proposed by authors such as Gould. Related to these three points, it also defends a nonoptimal, "non-struggling" view of evolution. Above all, by placing behavioral choices, shifts, and persistence at the very center, and as the primary drivers, of evolution—thus considering organisms to be key active players in their evolutionary history and the evolution of other organisms as well—it can account for the frequent occurrence of mismatches. If organisms as a whole were mainly passive players, and everything was "programmed" (gene-centered view) in the genome, decided by external forces (externalist view, e.g. by the external environment or a supernatural being), or related to vitalistic forces within the cells/atoms/tissues forming the organisms (vitalism), then in theory we should not expect mismatches to be so frequent.

Computers do not often make "bad choices" because they normally do not choose anything at all by themselves. However, organisms are active players that can potentially make an almost endless number of behavioral choices and, accordingly, of possible mistakes. Many of these choices are constrained by phenomena such as teaching/learning/imitating, thus leading to behavioral persistence, which can result over the long term in mismatches. There are also cases of behavioral choices that prove to be maladaptive from the very beginning. Behavioral evolution is therefore above all a trial-and-error process that allows, and often leads to, "bad choices"—except in those few organisms that have the capacity to undertake particularly complex, purposeful behaviors; however, even these organisms often make mistakes, as we clearly do (see, e.g., De Wall 2016).

In fact, two main points need to be stressed. The first concerns the concept of "nonoptimal'/'nonstruggling" and also more random view of evolution of ONCE. Within this view, it is entirely plausible that even for instance certain termite colonies that do not live in the "most" adequate places can still survive for long periods of time there, as long as there are enough resources in the region. This seems to be an inescapable fact considering that a huge number of termite colonies live in different geographical locations. Only one of these locations can theoretically be the "best"/most optimal one, but still they thrive in many others as well. As shown in the studies reviewed previously and stressed by many other examples provided throughout this book, far from optimal etho-ecological and/or eco-morphological matches actually are probably the rule, rather than the exception, on this planet. And this is very likely also related to the other major point: behavioral persistence. That is, even in the case of those termites that happen to choose the "most" appropriate place to make a nest at a certain moment in time, it is very likely that after a certain period of time and/or environmental change, that place would no longer be the "best" one. However, due to behavioral/ecological inheritance, it is likely that they would continue to live there, with such a scenario leading to non-optimal matches.

What is the "best" place for humans to live, taking into account our most common phenotypic traits? And why do all humans not live there? Why do some human groups persist in living in harsh environments—such as extremely cold and icy regions or extremely hot and dry regions—which in theory are not at all the most optimal places for us to live? Despite not being the most optimal places for us, at least some human groups have managed to live there, since long ago, generation after generation. The major reason for that is our behavior persistence. Taking this into account, and putting the organisms, particularly their trial-and-error behavioral choices and shifts and their behavioral persistence, at the very center of evolutionary biology helps to explain why etho-eco-morphological mismatches are so common on this planet.

Chapter 7
Internal Selection, Constraints, Contingency, Homology, Reversions, Atavisms, von Baer, Haeckel, and Alberch

A topic that is particularly crucial to ONCE is the occurrence and power of evolutionary constraints and how they can relate to evolutionary mismatches. My colleagues and me recently provided extensive discussions and several examples, in various recent papers and books, of internal constraints and internal selection [84, 88–90, 104, 106, 107, 112, 342]. Therefore, this chapter merely provides a short summary of the main points made in those publications that directly relate to the subject of the present work. Specifically, this summary is mainly based on—but also integrates a vast amount of new information not included in—Diogo et al.'s [106] review.

> **Box—History: Constraints, Structuralism, Physicalist Framework, Evo-Devo and "Deep Homology"**
> As explained in Chap. 6, eco-morphological and etho-morphological mismatches—such as the presence of pelvic bones in at least some stages of whale development—are examples that emphasize the importance of internal factors. These include internal constraints *sensu* Gould as well as *sensu* the physicalist framework, i.e. related not only to the conservation of developmental genetic/epigenetic mechanisms but to the physical proprieties of tissues, which result in an unfilled morphospace (see definitions and discussions in Chap. 1). They also include internal selection, as shown, for instance, by the death of human fetuses with more than 7 cervical vertebrae because of a general disturbance of the phylotypic stage, i.e. in cases in which the morphospace can be/is filled, but then internal factors lead to death in early stages of development (see Chap. 1 and later text). Authors such as Raff [302] divide internal constraints into "physical constraints" (e.g. there are only a few ways to form a tube), "genetic constraints" (e.g. related to genome size), and "developmental pleiotropy" (e.g. resistance of existing integrated organizations to changes or reorganization). He explains that current authors who tend to emphasize above all the "physical constraints" (e.g. within the

© Springer International Publishing AG 2017
R. Diogo, *Evolution Driven by Organismal Behavior*,
DOI 10.1007/978-3-319-47581-3_7

physicalist framework of researchers such as Newman/Müller) are "structuralists" who tend to see morphology as a result of the properties of physical rules that create structure in developing systems. According to him, these researchers are historically linked to a long-standing viewpoint on development that derives from pre-Darwinian idealistic morphology, which continued to be defended by authors who were opposed to Darwinism. For him, the main difference is that modern "structuralists" such as Newman and Müller are not opposed to Darwinian ideas.

An example of how structuralists were opposed to Darwin in the end of the nineteenth and beginning of the twentieth centuries is the book *Design in Nature* published by Pettigrew in 1908, which proposes that general rules of form are seen in both living and non-living systems (spiral galaxies *versus* animal spiral shells and horns). According to Raff, most "modern structuralists" share a "weaker version" of such structuralism and relate it with two current ideas. One, defended by authors such as Newman, is that general processes—e.g. "standard" physical processes including adhesion, surface tension, and phase separation—create general morphologies such as segments. The other, defended by authors such as Mittenthal, is that only a few morphogenetic processes and possible morphologies are consistent with particular biological functions. Raff argues that this modern "weaker version" of structuralism contrasts with the stronger one defended by some modern authors such as Goodwin [154, 155], who often use notions—e.g. a concept of homology based on shared morphogenetic principles instead of on historical continuity (see also [46])—that resemble those of pre-Darwinists.

Unfortunately, the increased power and increasingly gene-centered view of developmental biologists, and even of Evo-Devoists, within biological sciences are leading to similar confusions regarding long-standing definitions. For instance, increasingly more authors are using the term "homology" to refer to clear cases of phylogenetic parallelism (often called "deep homology" by many Evo-Devoists; see Miyashita and Diogo [260] for a recent review on this issue). Examples include the presence of complex eyes in invertebrates and vertebrates and, according to recent papers published by my colleagues and me, also the striking similarity of the forelimbs and hindlimbs of tetrapods. As explained in those recent papers, some genes/aspects of genetic networks are shared in the development of complex eyes in both invertebrates and vertebrates and of both the hindlimb and forelimb in tetrapods. However, the similarity of the complex eyes of vertebrates and invertebrates is clearly a derived (homoplasic) feature (because the last common ancestor of these two groups did not have complex eyes), as is the remarkable similarity of the hindlimb and forelimb within gnathostomes (vertebrates with jaws) [91–93, 95, 96, 104, 110, 260, 331]. That is, one must explain to which level one is referring: the genes involved in the formation of complex eyes in invertebrates and vertebrates might be homologous, but the eyes themselves, as a whole anatomical structure, cannot be because the last common ancestor of the two clades did not have such complex eyes.

Despite being aware of this contradiction, Raff [302: 353–355] states that both the leg fields in the fly *Drosophila* and limb fields in vertebrates express the Distal-less gene and that, in both, the hedgehog protein is related to a primary signal from posterior to anterior submodules, and then suggests that it may be parsimonious to infer "that the appendages have a deep underlying and completely unexpected homology." That is, he states that "appendages may have arisen from some primitive body projections in the common ancestor of vertebrates and arthropods, and these may have primitively required expression of Distal-less, hedgehog, and other shared regulatory genes." However, even if the last common ancestor of vertebrates and invertebrates had some "primitive body projections" that then led independently to the more complex appendages of insects and of vertebrates, these complex appendages would still not be homologous under the original (historical) definition of homology. This is because there is no historical continuity of the complex appendages themselves—i.e. of their phenotype, their form—since that last common ancestor.

As in the example of the complex eyes, one can say that some genes, or mechanisms, related to the development of those complex appendages might be homologous in vertebrates and invertebrates, but not the complex appendages as a whole anatomical unit. In fact, hindlimbs and forelimbs are derived anatomical structures only present in tetrapods within gnathostomes: neither of these limbs (i.e. including autopodia, or hands/feet) was present in the last common ancestor of the extant gnathostome taxa. That is, these are clearly examples of phylogenetic parallelism, or the so-called "deep homology"—probably related to developmental constraints but not to historical homology or historical serial homology [104, 260]. This was also pointed out by Minelli [258: 164], who stated that "animals provided with appendages were not necessarily derived from a common ancestor already endowed with appendages; the appendages of vertebrates are not the same as those of the arthropods, any suggestion from developmental genetics notwithstanding."

Hansen and Houle [173: 134] provided illuminating examples that illustrate the link between constraints and possible cases of evolutionary non-optimization in their interesting account of the relations between evolvability, stabilizing selection, constraints, and evolutionary stasis. One example concerns the wing shapes of dipteran insects (a clade including fruit flies). According to the authors, these shapes are under stabilizing selection. But they ask: "Are the optimal shapes likely to be nearly the same in thousands of species of widely different size, living under widely different conditions with respect to temperature, humidity, and wind conditions? Why does not allocation to wing mass and muscle depend on the relative importance of flight to energetic constraints? Should not shape depend on this allocation? Should not the importance of wings for mate choice have substantial effects on their optimal shape? Should not males and females with differently shaped bodies have wings more different in shape? And if there really is one global optimum that fits all these conditions, why then would thousands of similarly-sized hymenopterans (a clade comprising sawflies, wasps, bees, and ants) have such different wings?" As they point out, the main answer to these questions is that constraints clearly played a major role in the evolutionary history of dipterans in particular, as they do in evolution in general, and as is further illustrated by their example about body temperatures (see box below).

Box—Details: Puzzling Body and Testicular Temperatures, Mismatches, Pleiotropy, and Systems Biology
Another clear example of eco-morphological non-optimality provided by Hansen and Houle [173: 134] that further stresses that eco-morphological mismatches are very likely more frequent than commonly admitted concerns mammalian body temperatures. Almost all placental mammals keep their operating body temperatures between 37 and 38 °C and, puzzlingly, their testicular temperatures 1 °C below those temperatures. Therefore, they state, "if this is to be explained in terms of direct selection on body temperatures we need to show that an arctic lemming and an African elephant have similar ecologically determined temperature optima; this seems next to impossible in view of the huge differences in ambient temperature, heat exchange, metabolic needs and energetic constraints." Thus, they consider that mammalian body temperature seems unaccountably constant under a stabilizing-selection hypothesis. The explanation is, at least partially, associated to internal constraints—including e.g. pleiotropy, that is body temperature is related with interactions with other features—rather than exclusively to external conditions. This example thus is in line with my criticism, in Chap. 1, of Bonner's idea that larger/more complex organisms may be not only more affected by internal constraints—with which I would tend to agree—but also by external (natural) selection than smaller/less complex animals. That is, the example seems instead to conform more to a systems-biology framework, in which more complex organisms such as mammals are in general more homeostatic (less dependent on the external environment) than smaller and so-called 'less complex' organisms (see, e.g., [50]).

Young et al. [414] provided a further, powerful empirical example of internal selection (*versus* internal developmental constraints). Specifically, they showed that the period of reduced shape variance and convergent growth trajectories from prominence formation by way of fusion in the faces of amniotes, subsequently to which phenotypic diversity increases, is not due to developmental limits on variation. Instead, it is mainly related to selection against new trajectories that lead to maladaptive facial clefts. They further argued that the relatively high incidence of clefting in humans suggests that natural variation does occurs at this delicate ontogenetic time period but is then actively selected against. That is, the phenotypic convergence seen at this stage is the outcome of selection against genetic variations that would stimulate stage-inappropriate shape combinations and subsequent anatomical detrimental defects.

In my opinion, one of the papers that best reflects the highly constrained character of evolution and its ecological consequences was written by Futuyma [135: 1866]. He lists six main phenomena as examples of evolutionary constraints and "evidence of failure of adaptation". The first and most severe is extinction. The second concerns limits on the geographical and habitat ranges of species. The third is historical accident or contingency, as famously emphasized by Gould but "advanced earlier and repeatedly by Neo-Darwinists such as Mayr [249]" who stated that "if evolutionists have learned anything… it is that the origin of new taxa is largely a chance event." As noted by Futuyma, the origin of our lineage with our characteristic intellectual faculties was a unique event as were the origin of feathers, of the prehensile nose of elephants, and of the neural crest cells of vertebrates/chordates, and so on. The fact that many taxa are characterized by autapomorphies (i.e. unique features) is one of the most powerful illustrations that numerous evolutionary transitions are rare and exceptional, and thus very likely chance events.

The fourth phenomena listed by Futuyma is a consequence of the other three: there are "empty niches" as stressed by the occurrence of geographic and temporal disparities in the distribution of adaptive forms. The "unbalanced" biota of islands—even of large, ancient ones such as New Zealand—is one of his preferred examples of a more general condition. Fifth, for him "phylogenetic niche conservatism" is an ecological analogy to the also-conservative anatomical features that are often the synapomorphies of higher clades. As reviewed by Futuyma, niche conservatism is widespread in both plants and animals. For instance, numerous lineages of herbivorous insects have kept links with specific plant families for dozens of millions of years. The sixth phenomenon is the "stasis" in the fossil record—*sensu* Eldredge and Gould [132]—corroborated by endless studies reporting cases in which form is also kept without major changes for even longer periods of time. According to Futuyma, it is particularly striking—and not emphasized enough in the literature—that in many cases stasis does not refer to specific characters or anatomical regions but is instead seen in all the regions of the body as a whole.

An empirical study reviewed by Futuyama is particularly relevant to the evolution of overspecialization, discussed in Chap. 5. The study pointed out that the additive genetic variance for desiccation and cold resistance is notably lower in populations of 15 species of the fly *Drosophila* with limited tropical distributions than in 15 more broadly dispersed species. This example suggests that the "specialist" species lack genetic variation needed to expand their range/further adapt, as postulated by ONCE. Another illuminating study he reviews concerns a work in which researchers reciprocally planted families of a prairie legume (*Chamaecrista fasciculata*) from three latitudes in the three locations and calculated the selection gradient on various relevant features in order to evaluate how northern populations would change in response to a warming climate. They found additive genetic variance in each trait studied, but the expected rate of adaptation was slow because several genetic correlations among traits were opposed to the direction of selection on multiple traits taken concurrently.

Also of particular interest within the context of ONCE is the link made by Futuyma between constraints, behavioral persistence, and etho-morphological and eco-morphological mismatches. As noted by the author, community ecology is one of the areas that are more affected by the lack of understanding of the role played by constraints. Fortunately, this area is now being dramatically changed by the recognition that phylogenetic history is crucial to understand species composition and the diversity of communities. The major reason for this is phylogenetic niche conservatism (ecological inheritance *sensu* ONCE: see Fig. 1.2). For instance, numerous studies have shown that many of the species in a community did not evolve their features *in situ* but instead keep their ancestral ecological features when they dispersed from the ancestral region to their new habitats. For example, sclerophyllous leaves (hard leaves with short internodes) and other traits characteristic of shrubs in the California chaparral did not evolve *in situ*. The Californian species are instead members of a broadly distributed taxon that evolved these features elsewhere, much before California developed its current "Mediterranean" climate. Another example is that most host-specialized species of leaf beetles in New York State feed on the plant family used by their congeners in Europe or tropical America, and were thus likely able to disperse into/from northeastern North America only because those plant families were present in their new habitats.

> **Box—History: Goethe, the Romantic German School, Naturphilosophie, and Internalism**
>
> One of the earlier, more prominent defenders of internalism, who deeply influenced most of the researchers that defend internalist ideas or the authors that inspired them, was Goethe, the famous poet, novelist, playwright, and philosopher who created, among other notable works, the tragic play Faust. According to Goethe, internal forces (e.g. developmental ones) are the main sources of the phenotype, whereas the external environment mainly plays a secondary role of choosing between the restricted anatomical diversity shaped by these internal forces [308]. Interestingly, this idea is somewhat similar to that now defended by some proponents of the Extended Evolutionary Synthesis (Fig. 1.3). Goethe had a great impact on the Romantic German School (e.g. Oken) and Naturphilosophie (e.g. von Baer) and on Haeckel at the end of the nineteenth century as well as on non-German researchers such as Owen [307, 308] and Bateson [34]. In fact, Bateson compiled an impressive number of studies on animal morphology, human development, variations, and defects and defended ideas that are now becoming mainstream in Evo-Devo. For instance, he argued that variation is mainly due to internal mechanical (e.g. number of parts) or chemical (e.g. reactions leading to a certain color) factors (constraints) and that natural selection mainly selects between a very constrained number of phenotypes.

One of the most emblematic and extreme examples of an internalist view of evolution is Alberch's [7] ill-named theory, "the Logic of Monsters". According to this theory, which was based on a detailed skeletal study of digit reduction in amphibians [8, 9], there is often a parallel between the variation/defects in normal/ abnormal individuals of a certain taxon (e.g. humans) and the usual wildtype configuration seen in other taxa (e.g. lizards or amphibians). Such a parallel was also noted at the beginning of the nineteenth century by Meckel (1804), who stated that "the constant involvement of certain organs together in congenital malforma- tions allows the conclusion that their development is coordinated under normal conditions" [284]. This parallel is achieved through the regulation of a conserved developmental program (e.g. a set of genetic and/or epigenetic interactions) such that the structure of these internal interactions constrains the possible variation upon which selection can operate [7]. In principle, such internal constraints can break down in the evolution of some clades. According to Alberch, although in most clades this breakdown would lead to death of the embryos due to internal selection, members of other taxa might survive to adulthood.

A parallel between the more common phenotypic variations seen in the normal human population and malformations seen in birth defects is also to be expected according to Shapiro et al.'s [334] model, Lack of Homeostasis. This model was in large part formulated based on observations of human trisomic individuals and states that in such individuals the presence of a whole extra functioning chromo- some or of large chromosome segment causes a general disruption of evolved genetic balance. This results in decreased physiological and developmental buffering against genetic and environmental forces and to a general decrease in developmental and physiological homeostasis, in which more unstable pathways and processes are most often and most seriously affected, thus leading to variations in the normal population. An illustrative example, predicted by both the Logic of Monsters and Lack of Homeostasis models, is that a very common human poly- morphism—the absence of the palmaris longus muscle, displayed by approximately 15% of the normal population—is even more commonly found in humans with severe congenital malformations. This muscle was absent in 74% (105) of 141 defective upper limbs reviewed in Smith et al. [342].

Contrary to the lack of Homeostasis model, the Logic of Monsters predicts that defects are in general "logical" and "constrained" because constraints are generally maintained by internal homeostasis. That is why the Logic of Monsters predicts that congenital malformations and plastic variations found in a certain taxon often mirror features that are consistently found in individuals of other taxa. This pre- diction has been supported by studies showing that the existence of similar patterns between intra-specific diversity in a taxon (plasticity) and inter-specific diversity in different taxa are usually the result of similar developmental mechanisms [185]. The internalist framework of the Logic of Monsters thus contrasts with the more ex- ternalist view of adaptationists, who contend that the current form of organisms is mainly explained by the external environment in which they live and not by internal factors. For instance, frogs and salamanders tend to lose/reduce digits 1 and digit 5, respectively, due to developmental constraints: the first digit to be lost/reduced is the last to form in the development of each taxon. The reduction/loss of digit 1 is seen in frogs that live in very different environments and that are exposed to markedly different external factors [9]. Alberch and Gale explained these trends as the result of developmental mechanisms: the reduction of mesenchymal cells in the limb bud as a result of developmental truncation associated with dwarfism and a slow down in the rate of cell proliferation associated with paedomorphosis [275]. Similar examples concerning hand/foot skeletal elements in mammals have been provided recently by Senter and Moch [333]. Another example provided by Alberch [6] concerns St. Bernard dogs, which usually have an extra (6th) digit, probably because of their larger size and larger limb buds—smaller dogs of other species almost never have an extra digit and often even lack some digits—and not because the presence of a 6th digit is adaptive per se.

Many other examples of internal constraints are given in the literature including Arthur's *Biased Embryos and Evolution* [17], Minelli's *Forms of Becoming—The Evolutionary Biology of Development* [259] and *The Development of Animal Form* [258], Gould's *The Structure of Evolutionary Theory* [159], Blumberg's *Freaks of Nature—What Anomalies Tell Us About Development and Evolution*, and Leroi's *Mutants—On the Form, Varieties and Errors of the Human Body* and references therein. Minelli [258: 87] reviewed an example that I especially like because it is linked with complexity theory and the intrinsic limit of adaptive systems. Basically, the idea is that as the number of and interactions among parts increase, the number of conflicting constraints among the parts will also increase, thus suggesting that optimization can only attain increasingly poorer compromises/mismatches. That is, such limits cannot be overcome by stronger adaptive selection because the top limit depends on intrinsic structural rules. Minelli refers, as examples, to the apparent limit of three body elements such as the trimeous coelom (proto-, meso-, and metacoel; or axo-, hydro-, and somatocoel) of bryozoans (aquatic invertebrate animals often designated as moss animals), echinoderms (a clade including starfish, sea urchins and sea cucumbers) and other taxa, and of four body elements in flowers (sepals, petals, stamens and carpels) and in mammalian teeth (incisors, canines, premolars, molars).

Of course, as stressed by Minelli, number of kinds (related to "pure complexity" *sensu* [253]) is not the same as number of individual parts, which seems to be much less constrained (e.g., eunicids—a family of polychaetes, i.e. annelid worms—can have ≤ 1500 body segments). In derived vertebrates with >300 vertebrae such as eels and snakes that contradict "Williston's rule" (a general trend toward reduction in number of individual serial parts), the number of kinds remains the same as that seen in closely related groups with fewer number of individual vertebrae. Minelli suggests that if there is a trend in complexity, it is perhaps in the other sense: an increase in the number of body segments (i.e. polyisomerism) is often linked to a decrease in the complexity of the patterning along the main body axis as is the case in snakes, eels, centipedes, and millipedes according to him. That is, polyisomerism would in theory lead to a decrease of "pure complexity" *sensu* McShea, whereas anisomerism (loss/differentiation of individual parts and thus potential increase in number of kinds, in cases of differentiation) would likely often lead to an increase in "pure complexity". My colleagues and me recently discussed these concepts and compared the number of muscles and of different muscle types in various vertebrates in order to analyze the evolution of anisomerism/polyisomerism within these animals [108]. I will return to this topic later in the text.

In his outstanding book *Genetics, Paleontology and Macroevolution* Levinton [225] also stressed the importance of developmental (internal) constraints. He argued that these constraints are related to non-random (often non-continuous) canalization of evolutionary direction due to limits imposed by a complex interplay between gene expression and epigenetic interactions during ontogeny. He argued that the use of for instance Turing-like mechanisms (dynamic, unfixed mechanisms that are highly dependent on biomechanical signalling) during ontogeny frequently results in the development of a discrete number of complete structures: therefore, in a way ontogeny and thus evolution can be related to minor saltatory changes. Levington provided several animal case studies supporting a more internalist view of evolution and Alberch's Logic of Monsters but stressed that development is nevertheless probably more variable than Pere Alberch suggested. Arthur [15] reported additional examples of variable development, even within a genus, but noted that most of them concern non-vertebrate, particularly non-amniote, taxa. He wondered why, considering overall life cycles, are von Baer's laws (see later text) more pronounced in vertebrates than elsewhere? He answered that the high degree of embryo protection typical of vertebrates such as amniotes, and particularly of placentals such as humans, probably considerably reduces the strength of external selection pressures in early development. According to Arthur, although positive internal selection may be involved in the origin of phylum-level bodyplans, subsequent internal selection would thus likely be negative, thus taking the form of a "developmental constraint".

Box—Details: Constraints, Anatomical Networks, Modularity, Head Anatomy, and Birth Defects
The importance of internal constraints has been recently supported by the use of a new tool—anatomical network analyses—which may become key for comparative, developmental, and evolutionary biology. For instance, the reduction in number of skull bones is a macroevolutionary trend found across the ecologically diverse tetrapod clade (Williston's Law; see previous text) that seems to be at least partially related to internal biases favoring the loss of the least connected bones [125]. I will not describe in detail here the methodology employed in anatomical network analyses because that has been done in various recent papers by my colleagues and me [107, 126]. I think the example shown in Fig. 7.1 is sufficient to show the huge potential of this methodology and to illustrate an extreme case of developmental constraints that strongly supports Alberch's Logic of Monsters. Specifically, anatomical network analyses of the muscles and their contacts in the normal configuration of the adult human head have revealed that there are three main muscular modules: (1) an "ocular/upper face" module including facial muscles of both

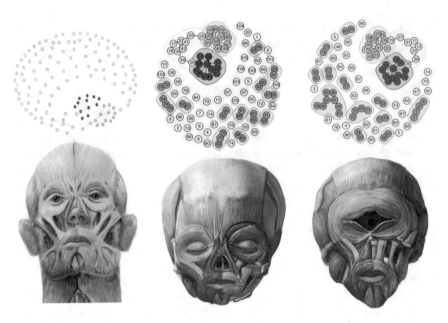

Fig. 7.1 Modules obtained from anatomical network analysis of head and neck muscles of the usual configuration seen in human adults (*left*) and newborns (*middle*) and of a cyclopic human fetus with trisomy 18 (*right*). Modified from Esteve-Altava et al. [126] and Smith et al. [342]. Note that, despite the severity of the cyclopic trisomic phenotype, the three muscular modules are exactly the same in all three conditions, i.e., there is always (1) an ocular/upper face module including facial muscles of both the left and right sides of the head, (2) a left orofacial module, and (3) a right orofacial module

the left and right sides of the head; (2) a "left orofacial" module including left facial muscles; and (3) a "right orofacial" module including right facial muscles (Fig. 7.1, left; [126]). The same modules are also found in the normal configuration of human newborns (Fig. 7.1, middle).

Strikingly, in an extreme case of congenital malformation seen in a cyclopic human fetus with trisomy 18 (described in detail by Smith et al. [342]), the very same three muscular modules are present despite the severity of the head defects and the cyclopia, thus showing that there is in fact a "logic" (order) even in such cases of extreme developmental deformities (Fig. 7.1, right). Anatomical, neurological, and pathological studies of humans further support the idea that these particular facial muscle modules are in fact deeply entrenched in the evolution, development, and overall organization of our heads. For instance, anatomical and innervation studies in humans and nonhuman primates suggest that the innervation of the face is bilaterally controlled for the upper part and mainly contralaterally controlled for the lower part and, accordingly, in humans paralysis of the upper face is often bilateral while that of the lower face is often unilateral [266]. This is in line with the facial-muscle modularity revealed by our network analyses (Fig. 7.1). Recent studies on modularity, for instance on the modular heterochrony of dermal *versus* endochondral bones, have also provided examples of strong internal constraints and pointed out how in many cases such constraints can play a crucial role in vertebrate macroevolution [212, 213].

The occurrence of internal constraints could also help to explain the results of our previous studies on the mode and tempo of primate and human evolution, specifically regarding the several examples in which rates of muscle evolution among various lineages of each major primate clade are strikingly similar [105]. According to a neutral model of evolution, this would be the expectation for molecular evolutionary changes. However, this had not been previously reported for any type of anatomical evolutionary changes, at least not within the order Primates. In this sense, these results support the importance of internal factors in primate evolution because, for instance, despite the major environmental and climate changes in Africa in the last 25 Ma, the rate of muscle changes accumulated during that period at the nodes that lead to the Cercopithecidae (the clade including all extant Old World monkeys) and subsequently to the genus *Colobus* (including colobuses) and also to *Cercopithecus* (including guenons) is exactly the same. Moreover, these similarities in global rates do not necessarily correspond to similarities in the rates for each different anatomical region. For example, at the node leading to the Cercopithecidae, the rate of head/neck changes is 0.19; at the node that leads to *Cercopithecus* it is 0.38; and at the node that leads to *Colobus* it is 0.00; the respective rates for the forelimb are 0.19, 0.00, and 0.38.

From an internalist view of evolution, these partial rates could be seen as support for the idea that ontogenetic constraints are so interconnected and strong that the potential for global change accumulated in the different body regions is limited. This fits with the results of studies showing that in early development, principally during the so-called "phylotypic stage", there is extensive interactivity among different modules of the body and therefore low effective modularity [139]. Furthermore, it has been argued that from a developmental viewpoint, if substantial somatic investment is made in one structure of a module of the body, this could limit investment devoted to the formation of another structure from that—or another —body module [139]. For instance, removal of the hindwing primordium on one side of the body of caterpillars results in an increase in weight of the adult butterfly forewing, thorax, and foreleg lying on that side, whereas the weight of these three latter structures does not change on the side where the hindwing primordium is intact [185]. It is possible that so-called "constructional trade-offs" constrain investment in whole phenotypes because the structural space in organisms is limiting [195]. Interestingly, Aristotle had similar ideas about internal constraints in the sense that he defended that a body part A may be limited by changes in a body part B [223]. Another empirical study supporting a similar idea is that of Narita and Kuratani [267], in which they refer to a puzzling "developmental trade-off" between the thoracic and lumbar vertebrae to maintain the sum of both at 19 in various mammalian groups as seen for instance in rats and wild boars (see Chap. 6).

Such examples of very stable muscle evolutionary rates within certain primate clades and of substantial rate differences between these clades also support Pere Alberch's idea that punctuated equilibrium might be related to events of long-term (constrained) stasis punctuated by periods of change (instability) due to the breaking of constraints [105]. This idea is somewhat similar to modern ideas relating minor *versus* major evolutionary transitions to changes in downstream genes (resulting in specific/generic differences) *versus* more internal portions (dissimilarities at the family, order, or class level) *versus* exceptionally conserved parts (differences at the phylum level) of genetic regulatory networks [69]. It is also somewhat related to De Beer's [72] suggestion that more stable evolution related to anagenesis (phyletic change) and to peramorphic events (terminal additions/developmental acceleration) would be punctuated by major changes in evolution and evolutionary rates related to cladogenesis (branching) and to neoteny (juvenilization, a form of paedomorphosis). This is because in paedomorphosis the deletion of terminal developmental stages would often lead to more generalized forms that may subsequently evolve in completely new ways. In contrast, in peramorphosis the addition of terminal stages would often result in very specialized forms adapted to very specific environments and therefore displaying relatively limited taxonomical diversity and low macroevolutionary potential (for more details, see [94]). This could in turn explain why empirical studies—such as those of McNamara [252]—inferring heterochronic changes of fossil animals from trilobites to vertebrates have found that, overall, paedomorphic and peramorphic processes have similar frequencies [302].

The definitions given in De Beer's [72] book *Embryos and Ancestors* were adapted and changed by Gould [156], who stated that developmental retardation leads to paedomorphosis (neoteny, or somatic retardation) and recapitulation (hypermorphosis, or retardation of maturation), as does developmental acceleration: it can lead to recapitulation by way of somatic acceleration or to paedomorphosis via progenesis (i.e. acceleration of maturation). Throughout the remainder of this book, when I refer to retardation and acceleration of development I refer to somatic ontogeny, i.e. acceleration leading to peramorphosis and specifically to recapitulation, and retardation leading to paedomorphosis and specifically to neoteny (Bolk's fetalization: see later test). However, it should be noted that development is very complex, and thus different parts of the same organism at the same time can have "different developmental ages" (*sensu* [258: 59]). This includes phenomena such as metamorphosis and heterochrony; for instance cockroach metamorphosis affects wings but not legs, and during the ontogeny of the butterfly *Precis coenia* there is a critical stage in which the animal is a larva and a pupa at the same time [258].

It should also be emphasized that not all evolutionary changes of ontogenetic trajectories occur by way of heterochrony. Some authors argue that heterotopy—changes in the place (as opposed to the time) of action of genes—could be even more important at a macroevolutionary scale (e.g. [327]). One example these authors provide refers to Carroll et al.'s [55] study on the fly *Drosophila*. The wings

Fig. 7.2 Paleozoic mayfly
nymph (*Kukalova
americana*). Modified from
Kukalová-Peck [214]

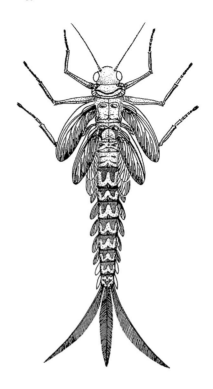

in pterygote (winged) insects originate from a segment of the leg, and fossil evidence suggests that early winged insects had wings on all thoracic and abdominal segments (see Fig. 7.2). Carroll et al.'s study showed that a series of homeotic genes repress wing development instead of promoting it; the phenotypic effect of knocking out these genes was the emergence of more wing primordia, thus showing how heterotypy—genes that are not expressed in some of the segments—can be related to a major macroevolutionary trend in invertebrates.

Box—Details: Phylotypic Stage, Hourlass Models, and von Baer's Laws of Development
Some authors argue that the evidence for a "phylotypic stage" and for a von Baer internalist evolutionary model is still scarce for non-vertebrate organisms (e.g. [37]), but it seems to apply better to amniotes and thus to mammals, including primates (e.g. [15, 225]). As discussed in Chap. 1, Galis [138] reported that changes in the stable number of seven cervical vertebrae seen in almost all mammals are nearly always associated with neural problems and an increased susceptibility to early childhood cancer and stillbirths in humans. This is probably caused by the breaking of developmental constraints: changes in the mechanisms (e.g. in Hox gene expression) leading to this number very likely perturb sensitive early stages of development, such as the

"phylotypic stage", thus leading to major abnormalities. Some recent studies claimed to provide evidence for a "molecular phylotypic stage" by showing a remarkably similar pattern of gene expression in early embryos of very diverse vertebrate clades (Elinson and Kezmoh 2010). Other "molecular hourglass" models have been recently proposed for plants [301], flies [273], and amniotes [197]. The existence of a phylotypic stage (or an hourglass model of development), either molecular or phenotypic, is a major problem for von Baer's law of divergence because divergence would apply only after this stage; in earlier stages development would be substantially variable and then mainly convergent [16, 18, 176].

Diogo and Wood [89] stressed that evolutionary reversions played a substantial role in primate/human evolution because 28 of the 220 evolutionary changes unambiguously optimized in the most accepted primate phylogenetic tree are reversions to a plesiomorphic state. Of those 28 reversions, 6 are directly related to our own evolution because they occurred at nodes that lead to the origin of modern humans, and 9 go against Dollo's law (which states that once a lost complex structure is unlikely to be regained). These studies support the idea that reacquisition in adults of morphological structures that were missing in adults for long periods of time is possible because the associated developmental pathways were kept in the members of that taxon. For instance, chimpanzees display a reversion of a synapomorphy of the Hominidae (great apes and modern humans) acquired at least 15.4 Ma ago, in which adult individuals have two muscles contrahentes digitorum—one going to digit 4 and the other to digit 5—other than the muscle adductor pollicis, which is the only contrahens muscle present in adults of other hominid taxa, including humans. Developmental studies of hand muscles [57] showed that karyotypically normal human embryos do have contrahentes going to various fingers, but these muscles are then usually reabsorbed or fuse with other structures during later embryonic development. Furthermore, in karyotypically abnormal humans, such as those with trisomies 13, 18, or 21, the contrahentes often persist as "atavisms" until well after birth [118]. Atavisms are coordinated and often incomplete structures that appear as developmental anomalies and resemble ancestral character states of the taxon to which the individual belongs [225]. Cihak [57] showed that the intermetacarpales are also present as discrete muscles in early embryonic stages of karyotypically normal modern humans before they fuse with some muscles flexores breves profundi to form the muscles interossei dorsales. Therefore, the evolutionary reversions resulting in the presence of contrahentes and discrete intermetacarpales in extant chimpanzees are likely related to heterochronic, specifically paedomorphic, events in the lineage leading to chimpanzees [89]. That is, in this respect, extant chimpanzees are seemingly more neotenic than humans. For recent works on atavisms concerning skeletal, instead of muscular, atavisms, see for instance Senter and Moch's [333] paper and references therein.

According to some authors, cases in which complex structures are formed early in ontogeny only to become lost/indistinct in later developmental stages (the so-called 'hidden variation') may grant organisms great ontogenetic potential early in development, so that if faced with external perturbations (e.g. climate change, habitat occupied by new species), evolution can use that potential (adaptive plasticity *sensu* [393]). However, as noted previously, according to authors such as Gould [156, 159] and Alberch [7], the occurrence of such cases supports a "constrained" (internalist) rather than an "adaptationist" (externalist) view of evolution. A major point of ONCE is centered on the notion of adaptive plasticity and hidden variation, which are crucial to allow the occurrence of completely new behaviors and then of morphological changes that are not necessarily related to genetic changes, such as epigenetic factors directly influenced by the environment (Fig. 1.2). Thus, in this respect, ONCE contrasts with the idea defended by Galis and Metz [139: 415–416], who stated: "without denying the evolutionary importance of phenotypic plasticity and genetic assimilation, we think that for the generation of macroevolutionary novelties the evidence for the impact of hidden variation is limited" (see also [225]).

The tendency to accumulate hidden variation could have been the outcome of the evolutionary success that plasticity confers in many situations. However, one surely cannot say that the persistence of individual structures—such as the muscle platysma cervicale in early human development (see Figs. 7.3, 7.4 and 7.5, and [141])—that then normally disappear during development is a phenomenon associated with

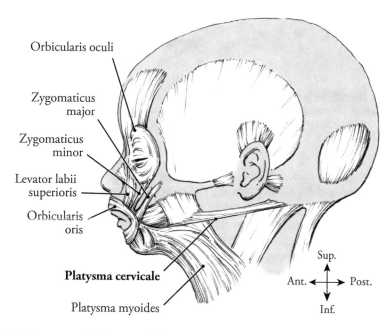

Fig. 7.3 Platysma cervicale, an atavistic muscle sometimes present in modern humans, here originating from the posterior platysma myoides and inserting with the sternocleidomastoideus onto the mastoid process. Anomalies are labeled in bold. Modified from Smith et al. [342]

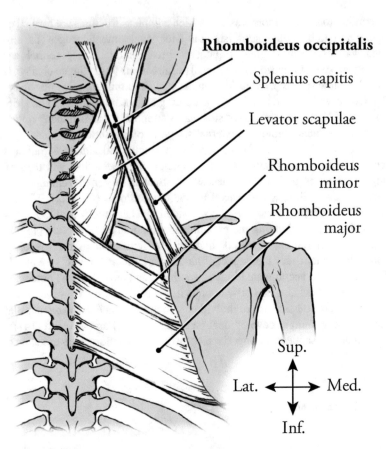

Fig. 7.4 The rhomboideus occipitalis, an atavistic muscle sometimes present in modern humans, here spanning from the superior angle of the scapula between the rhomboideus minor and levator scapulae to the occipital region of the skull. Anomalies are labeled in bold. Modified from Smith et al. [342]

specific positive selective forces. Nature cannot guess the future and thus cannot predict whether the muscle platysma cervicale might be something that will be eventually later "useful" in the future of humanity. Instead, the platysma cervicale is kept during early human ontogeny mainly due to evolutionary/developmental constraints, and the occurrence of such constraints is what may lead in the future to an eventual phylogenetic reversion in which adult humans will normally have that muscle be it by chance (random factors, such as neutral random mutations) or because this might lead to an advantage linked to a specific behavioral/ecological shift. The release of hidden variation—usually related to developmental/phylogenetic constraints—is likely a common event during changes from stabilizing to directional selection [327]. In fact, apart from phylogenetic reversions, there are other types of reversions, such as the so-called 'Lazarus developmental reversions' *sensu* Minelli

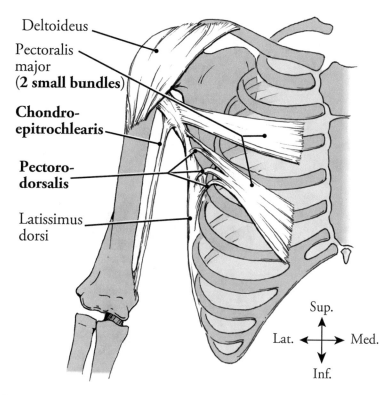

Deltoideus

Pectoralis
major
(2 small bundles)

**Chondro-
epitrochlearis**

**Pectoro-
dorsalis**

Latissimus
dorsi

Sup.

Lat. ◄──┼──► Med.

Inf.

Fig. 7.5 Dorsoepitrochlearis, or chondroepitrochlearis, an atavistic muscle sometimes present in modern humans, here spanning from the latissimus dorsi to the medial epicondyle of the humerus. Anomalies are labeled in bold. Modified from Smith et al. [342]

[258: 75]: features that disappear from an organism's body at a given ontogenetic stage and then reappear at a later stage. An example given by Minelli concerns the fourth pair of legs in mites (small anthropod invertebrates closely related to ticks): the legs are present in the embryo, lacking in the larval instar, and then present in the nymph and adult stages of most mites.

Importantly, the platysma cervicale, contrahentes digitorum, and inter-metacarpales of karyotypically 'normal' human embryos do not correspond to the muscles of adult primates such as chimpanzees or of other adult mammals: they correspond instead to the muscles of the embryos of those taxa [88, 89]. The developmental pathways resulting in the presence of these muscles in adults of those taxa were not completely lost in modern humans, even after several million years, likely because these pathways are associated (pleiotropy) with those recruited in the formation of other structures that are present and functional in modern human adults. Examples of abnormal human "atavisms" that were historically used to support Haeckel's recapitulation theory and thus the importance of peramorphosis in human evolution [70, 399] were summarized and criticized by De Beer [72] and more recently by Verhulst [366]. For instance, Darwin and many other nineteenth

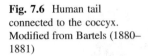

Fig. 7.6 Human tail
connected to the coccyx.
Modified from Bartels (1880–
1881)

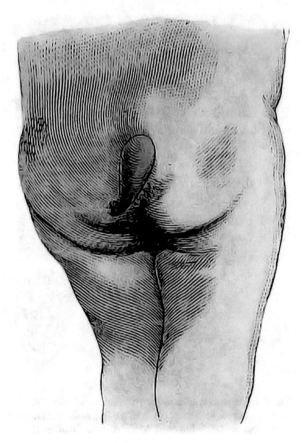

century authors considered the presence in some human newborns of a tail-like
appendage at the height of the coccyx or lumbar spine (Fig. 7.6) to be an "atavism".
Verhulst [366] attacked this idea by pointing out that these appendages are very
different from animal tails, e.g. they almost never contain bones (vertebrae) or
cartilage. However, it should be noted that some authors have described muscles
associated with these tail-like structures that, according to them, do resemble caudal
muscles of other animals [399].

In fact, it is now becoming clear that many of the human "tails" described so far
are true tails, as recently noted by Tubbs et al. [355]. During weeks 4–6, human
embryos have tails with 10–12 caudal vertebrae and a distal portion that has
mesodermal elements and lacks bone. By the end of the week 5, the tail reaches its
maximum relative length (approximately 1/6 of the length of the whole embryo).
By week 8, the tail normally disappears; the distal vertebral (coccygeal) segments
become phagocytosed by white blood cells, whereas the surrounding vertebrae
become reduced, and the cells of the distal mesodermal non-bony portion are also
phagocytosed. However, in a few humans an appendage persists until later

developmental stages, including adulthood, as a result of the disturbance of the normal deterioration of the true embryonic tail. Remarkably, reflex (e.g. associated with crying or coughing) or even voluntary (associated with the presence of voluntary muscle) movement of such tails has been reported in some instances. If we remember that tails disappeared phylogenetically >10 million years ago in the evolutionary history of our lineage and that they are not present in adults of any hominoid (humans + apes) extant taxon, the postnatal presence of such true tails in a few humans is truly impressive and reminds us again how profoundly constrained evolution is.

Some muscle variants in the normal human population or in humans with congenital malformations are also truly atavistic as explained previously (e.g. platysma cervicale, rhomboideus occipitalis, opponens hallucis, dorsoepitrochlearis or chondro-epitrochlearis, epitrochleoanconeus, levator claviculae: Figs. 7.3, 7.4, 7.5). It is not clear whether or not the presence of these atavistic muscles is due to a developmental delay. More detailed studies on the ontogeny of these structures in humans and other animals, particularly primates, are needed to clarify these issues. Hall [167] reviewed some potential examples of atavistic features in humans and other animals and suggested that the presence of such features as human variants might represent maintenance of a polymorphism in the normal human population. For instance, as noted previously, the muscle palmaris longus is present in approximately 85% of the normal human population: as in most primate taxa this muscle is present in approximately 100% of the normal population, there was a trend toward a decreased frequency of this muscle in human evolutionary history [88, 89]. If this trends continues to occur, it is likely that in the future this muscle may be present in only a very small percentage of the normal adult human population and would thus be seen as a rare atavistic variation/anomaly. That is why although atavistic features are often related to developmental arrest/delay and thus to the process that leads to paedomorphosis, they are often used by authors defending the importance of the opposite phenomenon, i.e., peramorphosis and recapitulation. However, it should also be noted that, contrary to many of the examples provided previously, numerous so-called "atavisms" described by previous authors, for instance in trisomic humans in the 1970s through 1990a (e.g., [28, 118]), cannot be true atavisms by definition because those features were never present in our direct ancestors [88–90, 342].

Box—History: Bolk's Fetalization Theory, Haeckel's Recapitulation, Neoteny, Apes, and Human Anatomy
Some authors use cases of abnormal development in which paedomorphosis leads to true atavisms to support the theory that the opposite pattern occurred in human evolution, i.e. that those atavistic structures were once normally seen in adults but then due to peramorphosis became only present in earlier stages (as e.g. the platysma cervicale) of, or even completely absent in, ontogeny. The followers of Bolk's [43] fetalization theory (i.e. the idea that humans are mainly neotenic apes, which was partially accepted by Gould

[156]) aimed instead to find cases of paedomorphosis in normal human development to support the idea that this phenomenon was important for the evolution of normal human anatomy. The example of the muscle palmaris longus also stresses an important point made by Holmes [191], i.e. that, although it is not common, it is likely that there are some examples of Haeckel Recapitulation in the literal sense. In other words, because the late fetal and adult configuration of the human palmaris longus are essentially the same [29, 30], if in the future this muscle is present until late fetal stages but never in adult stages of the normal human population, the fetal stage will in fact be similar to the adult ancestral stage.

Many of the examples provided by Verhulst [366] to support Bolk's fetalization theory are clearly as flawed as many of the erroneous "atavistic" examples provided in the past to support the opposite (recapitulation) theory. For example, in a desperate attempt to fit the very complex and derived human hand into Bolk's fetalization theory, Verhulst argues that the configuration of the human hand actually represents the ancestral, fetal stage in the sense that gorillas subsequently lost (in evolution/development) the ability to use the precision grip. Although it is clear that our hands also have various plesiomorphic (more ancestral) features (e.g. [12]), embryological and morphological studies (e.g., [89]), as well as recent genetic studies, clearly support the idea that they also display a wide range of derived characters. For instance, Prabhakar et al. [297] suggested that there was a gain of function in one of the most rapidly evolving human noncoding elements (HACNS1) in the human hand, which likely altered the expression of nearby genes during limb development (although it should be noted that some aspects of Prabhakar et al.'s study are still controversial: Terence Capellini, personal communication).

However, it should be emphasized that humans do seem to have neotenic features. For example, it is accepted that the skulls of apes and humans closely resemble one another early in development, and that subsequently the facial region develops more slowly in humans, thus retaining the proportions of juvenile apes [53]. Some detailed comparisons regarding other features, such as the neuronal system, have also supported the idea that human brain development is retarded compared with that of other primates, particularly in some association areas related to episodic memory, social navigation, and planning, thus allowing us to retain juvenile characteristics/learning over a long lifetime (neoteny: [48]). This neotenic ability to retain such juvenile features over a long lifetime due to retarded development seems to be combined with an accelerated prenatal development of the brain in humans [322], therefore emphasizing the point that the same taxon can display a mosaic of retarded/accelerated features even within the same organ/anatomical region.

The majority of these and other examples provided in the literature regarding differences between the ontogeny of humans and other primates, either to support neotenic or peramorphic views of human evolution, show a similar pattern: almost all refer to examples of terminal addition/acceleration (leading to peramorphosis) or deletion/retardation (leading to paedomorphosis). That is, these examples seem to support Levinton's [225] view that changes in development are mostly terminal. This accords with the idea that earlier developmental stages in vertebrates are particularly sensitive and/or prone to internal constraints (e.g. "phylotypic stage": see Holland [190] and previous text). Curiously, some examples are used by various authors to support Haeckel Recapitulation and by others to support the opposite idea, i.e. Bolk's human fetalization. For instance, the supposed resemblance of some so-called "hairy humans" to other primates has been interpreted by some as an atavism that would support Haeckel's view of evolution. However, as stressed by Leroi [222, p. 285], "both the hairy Burmese and Canary Islanders are described as having exceptionally fine, silken hair; this does not really resemble the robust pelt that covers adult apes" but instead the "fine, silky", "lanugo" hair of human fetuses. The adult abnormality is thus probably due to a developmental arrest/delay, and emphasizes again the major error of Haeckel's recapitulation theory: the hair of human fetuses does not resemble the hair of adult apes, being instead probably more similar to the hair of ape fetuses.

In fact, hair itself was also used as an example of human neoteny (paedomorphosis) to support Bolk's "fetalization theory" in the sense that newborn gorillas have hair on the head and short, "lanugo"-like hair in the rest of the body; their characteristic hairy coat does not appear until later stages [180]. In this case, human adults would seem to be neotenic. However, with respect to hair, humans are both neotenic and peramorphic in the sense that they not only not form a hairy coat (terminal deletion of an ancestral character) but also lose even the "lanugo" body hair during normal development (terminal addition of a special, derived character). This is a further example of a parallel between development (humans lose "lanugo" body hair during development) and evolution (human adults lost most body hair during evolution) but not of human neoteny (paedomorphosis) or recapitulation (peramorphosis).

In summary, humans seem to have a mosaic of both paedomorphic and peramorphic features, and only a more detailed, careful, systematic, and less tendentious study comparing normal and abnormal human development with the development/adult anatomy of other animals can quantify the specific contribution of each of these two events during human evolution. Among the 1540 cases of human muscle congenital malformations compiled by my colleagues and me [342], 257 (17%) are potential atavisms and 352 (23%) may be due to developmental delay. Therefore, if more detailed comparative developmental studies confirm that a substantial portion of these cases are truly atavistic and/or due to developmental delay, they would represent a high proportion of the total number of malformations. However, these phenomena would still represent a minority of all cases of birth defects we have compiled. This supports the results of other studies of trisomic human individuals suggesting that, apart from developmental retardation, many other factors are probably involved in the anatomical defects in these individuals, including the aberrant organization of primordia (e.g. presence of two rather than three pulmonic valve cups) and misdirected morphogenetic movements (e.g. leading to abnormal innervation of muscles or "horseshoe kidney": e.g., [28]).

Box—History: Haeckel's Recapitulation, Racism, von Baer's Ideas, Progress, and the Scala Naturae

The examples of the frequent occurrence of atavisms, in particular of evolutionary reversions, given in the previous paragraphs and boxes, especially those about human evolution, go against the notion of progress and purpose of evolution that was closely related to the rise of developmental theories such as Haeckel Recapitulation ([156]; Gould 1981). This theory was based on the "scala naturae" view of nature in which white human males were seen as more complex and therefore placed at the top of the scale. This scheme had profound and extremely unfortunate social and racial implications (Gould 1981) and, unfortunately, continues to be deeply embedded in some current textbooks and scientific papers [83, 108]. Haeckel recapitulation theory is no longer accepted; in fact, one of von Baer's four laws—which are now commonly accepted in the literature—directly goes against it: the law stating that the ontogeny of an animal does not, in general, recapitulate the adult stages of its ancestors. The other three laws of von Baer are: (1) general features of the embryo appear earlier than special features; (2) special characters develop from general characters through the increase of tissue and organ differentiation; and (3) embryos of different species progressively diverge from one another during ontogeny [2].

However, Haeckel, like Lamarck, is often used as a "straw-man" in biological sciences, in particular in evolutionary biology, despite the fact that each of them also had great achievements within these scientific areas (see, e.g., [307–309]). Specifically, researchers often use Haeckel as a "straw man" to deny that there is often a parallel between phylogeny and development. According to Gould [156] such a parallel does however exist, and is probably driven more by phylogenetic/ontogenetic constraints than by adaptive plasticity as noted previously. This view was also supported by Levinton [225: 219–220] based on his compilation of a large amount of data from both extant and fossil animals. He compiled a list of examples provided by De Beer [72] to contradict Haeckel Recapitulation, e.g. "teeth evolved before tongues, yet tongues appear before teeth in mammalian development", but then stated that a review of the data available suggests that ontogeny and phylogeny are in fact often intimately related because evolution usually involves terminal additions (*contra* [239]; see later text). Levinton argued that Haeckel

Recapitulation can in fact be considered a special case of von Baer's law in which evolutionary terminal additions play a central role. Levinton therefore formulated his own, more specific ontogenetic-phylogenetic laws as follows: (1) many major structures have some kind of integration via development; (2) the ontogeny of a structure involves an order of appearance of its sub-structures (e.g. proximo-distal elaboration of the tetrapod limb); (3) when an order exists, evolutionary changes tend to favor later ontogenetic stages first; and (4) later ontogenetic phases, in general, are likely of "lower burden" and thus more likely to be modified.

My previous studies with other colleagues do show that in the case of the muscles of zebrafish, salamanders, and frogs, the order in which muscles appear in ontogeny is usually similar to the order in which they appeared in phylogeny [92, 98, 100, 106, 107, 409, 410]. The parallelism between phylogeny and development is also supported by studies on human evolution and ontogeny. For instance, the flexores breves profundi muscles normally appear in 13.5-mm (total length) human embryos, the intermetacarpales, lumbricales, contrahentes, and dorsometacarpales at 14 mm, the abductor pollicis brevis at 15 mm, and the abductor digiti minimi at 16 mm [57]: all of these structures were acquired phylogenetically before the origin of mammals [87]. The adductor pollicis, opponens digiti minimi, and flexor digiti minimi brevis appear in human ontogeny before 20 mm: these structures are found in mammals such as monotremes and rats. The opponens pollicis and flexor pollicis brevis appear at 28 mm: these muscles are only consistently found in primates. Then, the contrahentes become completely undistinguishable at 35 mm: these muscles were phylogenetically lost in the node leading to hominids (i.e. great apes and humans).

Importantly, the reabsorption/fusion/loss of structures such as the contrahentes, as well as the intermetacarpales or platysma cervicale or the "lanugo" hair during normal human development, does not fully match the expectation of von Baer's second law. As noted previously, according to this law a more general "mammalian" configuration (e.g. having body hair or contrahentes) should be followed by an increasing tissue and organ differentiation, i.e. the development of specialized, more derived structures from the general configuration [2]. Von Baer stressed (*contra* some supporters of Haeckel Recapitulation) that there is no "trend towards perfection" during human development, but stated that there is a "trend towards complexity" in development [156: 55]. However, the examples provided above show that in reality there are many cases in which there is instead a simplification/loss of structures during human ontogeny. Other similar examples concerning human embryonic development are well known: for instance, the morphogenesis of five pairs of aortic arches (which once sent blood to the gills) is followed by complete destruction of two of them, during human ontogeny. As Held [180: 67] put it, this "only makes sense as a historical constraint: it must have been... easier to reconfigure the existing plumbing than to scrap it altogether and start afresh."

Structures are also formed and subsequently lost during the development of many other chordate taxa we have studied. For example, in neotenic salamander species such as axolotls that do not undergo full metamorphosis, some muscles become indistinct/lost/reabsorbed during ontogeny, e.g. the pseudotemporalis profundus and levator hyoideus become integrated in the pseudotemporalis superficialis and the depressor mandibulae, respectively [409]. Another example concerns the notochord, which is present in adult stages of phylogenetically basal chordates *versus* only in the early developmental stages of vertebrates. This is probably due to the fact that the physical presence, as well as the signals emanating from the notochord, are essential prerequisites for differentiation and morphogenesis of the spinal cord, vertebrae, and gut tube structures [149, 150, 191]. In other words, the presence and subsequent loss of the notochord is evidence of evolutionary tinkering [200]. Of course, there are also numerous examples of muscle differentiation in human ontogeny, e.g. the rhomboideus complex becomes divided into a rhomboideus major and a rhomboideus minor, and the extensor carpi radialis divides into brevis and longus muscles [29, 227]. Importantly, these latter examples that conform more to von Baer's second law also support the parallelism between phylogeny and development because ancestral adult tetrapods first had an undifferentiated rhomboideus and an undifferentiated extensor carpi radialis, and each of these two muscles became subdivided only later in tetrapod evolution [88].

Another major problem with von Baer's theory, as discussed by some authors in the past but unfortunately largely neglected nowadays, is the lack of a dynamic, evolutionary perspective [341]. Von Baer was an antievolutionist: his most famous works were done in the 1820s before the publication of Darwin's books and were mainly against "romantic evolution", but by the end of his life he did publicly oppose Darwin. Therefore, von Baer's rules assume rigidity over time, while in reality the "general" and the "specific" in evolutionary morphogenesis are continuously in motion: the general changes to the specific and the specific to the general, and both categories experience ceaseless changes [341]. For instance, birds never develop teeth during normal development: the general tetrapod (or gnathostome) configuration—having teeth—is no longer found at all. Avian neural crest cells are competent to induce teeth, but the avian oral epithelium has lost the capability to respond properly, despite the presence of the ancestral coding region for enamel in modern birds [211]; for a more recent work on the subject and a description of generic correlates, see [257]). As a result, the embryos of birds and other reptiles do not "keep up" with each other during the earlier ontogenetic stages, and a feature that was surely present in avian ancestors is completely missing in normal bird ontogeny. This example thus contradicts even a "von Baerian" type of recapitulation, i.e. where the embryo of the offspring should be identical or resemble the embryo of the ancestors and not the adult animal as stated in Haeckel Recapitulation [237, 341]. That is why De Beer [72] proposed the term "repetition" (ontogeny repeats phylogeny) and Holmes [191] speaks more of "parallelism", a term similar to the "Phylo-Devo parallelism" expression used in my own recent works (see text below).

In the paper from which some of the paragraphs of this chapter are based, my colleagues and I formally defined the term "Phylo-Devo parallelism" to refer to the common parallelism between phylogeny and evolution [106]. This term is crucial for ONCE because it emphasizes the crucial role played by developmental constraints in evolution (see previous box). We use this term specifically to avoid confusion with "evolutionary parallelism", which refers to the resemblance between the evolutionary history of two different taxa and not between evolution and development. Phylo-Devo parallelism thus refers to the fact that in most cases of phenotypic change occurring in the normal development of a certain taxon, the order of the developmental changes is similar to the order of evolutionary (phylogenetic) changes that occurred during the evolutionary history of that taxon. This notion is different from Haeckel Recapitulation because there is no necessary reference to adult forms, i.e. the parallelism applies both to adult (in the few cases where there is a literal Haeckel Recapitulation and non-adult stages (see previous box and text). Phylo-Devo parallelism is also different from von Baerian recapitulation (*sensu* [237]) because not all stages seen in the embryos of the ancestors are necessarily "recapitulated" (e.g. teeth are never seen in bird development). Moreover, it is also different from von Baer's laws because it provides the dynamic evolutionary frame stressed by Slaby [341] and thus removes the rigidity/immovability about the general, the specific, and the supposed trend toward increasing tissue and organ differentiation from the general condition to the specialized, more derived structures (see previous box). Phylo-Devo parallelism can thus account for atavisms, i.e. complex structures that were once generally present in adults but are now almost completely suppressed in development. As noted previously, an example is the platysma cervicale (Fig. 7.3), which is formed and then disappears in normal human development, thus mirroring its appearance in the mammalian clade and disappearance in our more recent evolutionary history. Many other examples of Phylo-Devo parallelism have been mentioned previously, as well as in many works of other authors about several different groups of organisms, one of my favorite ones concerning the evolution of a remarkable termite defensive organ (see [205]).

However, it should be noted that although Phylo-Devo parallelism appears to be the rule in most of the numerous vertebrate taxa for which we have compiled musculoskeletal data so far (see previous box), there are a few exceptions. For instance, among the 58 head muscles of zebrafish, salamanders and turtles for which we explicitly compared "phylo *versus* devo" data, five (8.6%) break this rule (2 out of the 29 muscles examined in the zebrafish, 2 out of 23 muscles examined in the axolotl, and 1 out of the 6 muscles examined in the turtle *Trachemys*: [85, 99,

409]; see also box above. Whereas the cases conforming to the Phylo-Devo parallelism rule are very likely due to internal constraints as explained previously, a substantial proportion of the exceptions probably relate to direct adaptations to specific external conditions. For instance, one of the five exceptions is that, in 4-day zebrafish larvae, the levator arcus branchialis 5 is already much more developed than other branchial muscles, before the mandibular (first arch) muscle adductor mandibulae splits into bundles. In contrast, in evolution hypertrophy of the levator arcus branchialis 5 occurred in the node leading to the Cypriniformes (teleost fishes including carps) evolutionarily much later than the division of the adductor mandibulae into bundles [99]. The alteration of the levator arcus branchialis 5 and of the skeletal element that is moved by it, the ceratobranchial 5, is related to the specialized feeding mechanisms of Cypriniformes, in which ceratobranchial 5 bears teeth and ossifies earlier than other ceratobranchials—a case of developmental acceleration. Such coordinated developmental timing changes may therefore ensure proper size relationships between muscles and skeletal structures [99]. This suggests that cases of Phylo-Devo parallelism are probably related to the idea that most major phenotypic changes occur later in development due to internal constraints, whereas exceptions to this rule are related to earlier changes that are likely related to the breaking of those constraints and that, if viable, have a high adaptive/cladogenic potential. In this case, the peculiar acceleration of development of the levator arcus branchialis 5 and ceratobranchial 5 in cypriniforms is related to a peculiar synapomorphy of these fishes—the acquisition of a new, pharyngeal jaw—that very likely contributed to their great taxonomic diversity (see [82]).

Exceptions to Phylo-Devo parallelism are also expected in cases in which markedly different developmental processes lead to similar adult configurations, for instance due to canalization (buffering of a phenotypic character against variation in the ontogenetic mechanisms related to its development: Rice [306]; see Chap. 8). In some of those cases, the developmental differences can be attributed to selection acting directly on early ontogenetic stages associated, for example, with life-history changes such as a shift from indirect to direct development. Janine Ziermann and I recently provided examples of changes in the order of muscle appearance in the heads of frogs with direct *versus* indirect development, which have a very similar adult configuration [410]. However, there are also major changes in early developmental stages that do not appear to be under direct selection and continue to produce a similar adult morphology [306]. As explained previously, the order of formation of the hand digits in salamanders is the opposite of that seen in frogs (radio-ulnar *versus* ulno-radial, respectively), mainly due to internal constraints. However, due to other types of internal constraints (e.g. leading to a consistent pattern of topological muscle–bone spatial associations: see Chap. 8), the overall adult muscle configuration of the hands of frogs is generally similar to that seen in salamanders [92, 93]. For a few other examples of exceptions to Phylo-Devo parallelism, see for instance Weisbecker et al. [392], Koyabu and Son [212], and Smith [343].

The fact that most authors, including von Baer, tend to neglect the common occurrence of secondary loss/simplification of structures during development is very likely related to a strategy that is often followed in developmental biology:

idealization [235]. That is, developmental biologists tend to see phenotypic variation and secondary loss/simplification during the development of wildtype model organisms, including humans, as noise. Examples of this propensity include the very simplified, and erroneous, descriptions of muscle development in humans as a trend toward progressive structural differentiation (see review of Diogo et al. [112] and previous text). Another example is the establishment of "normal stages" of development of model organisms under strict laboratory conditions in order to deliberately reduce developmental plasticity and thus to have "standardized" comparisons of their "normal" phenotype between different laboratories around the globe [235]. As a self-criticism, I can thus admit that, although in my previous works I tried to account for variations in both normal and abnormal development, particularly in humans as stressed previously, I mainly emphasized examples favoring a more internalist view of evolution. That is, I myself was partially influenced by the fact that most data available on model organisms were obtained under such "idealized" conditions. However, these data do not reflect the true developmental plasticity and behavioral richness of organisms. As explained in the preface, this book is an attempt to complement my former works and integrate data stressing both this richness, on the one hand, and the importance of constraints in evolution, on the other hand.

Chapter 8
ONCE Links Internal Factors, Epigenetics, Matsuda, Waddington, Goldschmidt, and Macroevolution

It is in interesting to see how increasingly more researchers—particularly the promoters of an Extended Evolutionary Synthesis (Fig. 1.3) within the context of the data recently obtained in Evo-Devo, epigenetics, niche construction and ecological studies—are promoting the work of a previously little-known but fascinating author: Ryuichi Matsuda. One of those researchers is the always-informed Brian Hall, who co-edited a book that in a way was also a tribute to Matsuda: *Environment, Development and Evolution—Toward a Synthesis* [172]. Hall noted that Matsuda linked environment (e.g., settlement on the host), shifts in endocrine balance, phenotypic modification, and evolutionary transitions and was particularly aware of the relations, as well as subtle differences, between reaction norms, organic selection, and genetic assimilation [170: 152–154]. A crucial element of Matsuda's approach was that adjustments in organismal physiology, including endocrine status, are the initial response to environmental shifts, thus "explaining" the instant response of living beings to these shifts. Therefore, although the physiological–endocrinological change is not inherited, for Matsuda the ability to elicit the changes in response to the new habitats is itself heritable and therefore individuals continue to exhibit such adaptive changes over generations. However, a point that is often neglected—and that is discussed throughout the present book—is that within this scenario similar changes are an answer to similar habitats: therefore, this does not explain the common occurrence of long-term evolutionary trends that span for several millions of year and across very different habitats.

A series of recent studies by Shkil, Smirnov, and colleagues (reviewed in [336]) have provided further support for the idea that epigenetic hormonal changes can result in developmental shifts that lead to drastic morphological differences that might, however, have been crucial for other macroevolutionary events. This includes the evolutionary history of serially similar structures and of polyisomerism and anisomerism (see Chap. 8; for a more detailed, recent review, see [108]). When Shkil and co-authors increased the level of thyroid hormone in specimens of vertebrate groups, such as zebrafish and amphibians, the number of serial structures—such as pharyngeal teeth, branchiostegal rays, fin radials and rays, and scales—often

© Springer International Publishing AG 2017

R. Diogo, *Evolution Driven by Organismal Behavior*,
DOI 10.1007/978-3-319-47581-3_8

decreased (anisomerism). According to the authors, this was due to an acceleration, and precocious stop, of the development of these structures. In contrast, when they decreased the level of thyroid hormone, the number of serial structures often increased (polyisomerism). According to the authors, this occurred because the development of these structures was decelerated and thus prolonged. Interestingly, the decrease in thyroid hormone led to an augmentation not only in the number of these serial structures but also in non-serial structures such as ossification centers of the pectoral girdle of the zebrafish and of the skull of amphibians.

One might speculate that these results indicate that the early evolution of serial structures in taxa such as the vertebrates, in which the number of serial structures tends to be ancestrally higher (polyisomerism) could be related, at least partially, to a decrease in the level of hormones such as the thyroid ones. One could therefore relate such a decrease in hormone level to a deceleration of development, e.g. to neoteny according to Shkil and Smirnov [336], thus relating neoteny to major macroevolutionary innovations/changes and the origin of higher clades as suggested by Gould [156, 159]. In contrast, one could try to connect the (supposed) subsequent decrease in serial structures in more derived vertebrate taxa—by e.g. loss or transformation of some of these structures (anisomerism)—with an increase in thyroid hormones and thus an acceleration of development, e.g. with peramorphosis according to Shkil and Smirnov. In this way, one could connect the evolution of serial structures within each higher clade to a more frequent occurrence of peramorphosis (as opposed to the origin and early evolution of different higher clades more related to neoteny) as proposed by authors such as Haeckel (Chap. 7; see also Diogo et al. [106] for a recent review). However, the statement of Shkil and Smirnov [336] that acceleration of development leading to a decrease in serial structures can result both from a somatic (peramorphosis) and sexual (progenesis) acceleration, as well as that an increase in serial structures can result from both a somatic (neoteny) and sexual acceleration, makes it difficult to connect with these broader ideas and the evolutionary theories of authors such as Gould or Haeckel.

Box—History: Matsuda's Pan-Environmentalism, Étienne Geoffroy St. Hilaire, and Lamarck

As noted previously, the book edited by Hall et al. [172] was in a great part dedicated to Matsuda's pan-environmentalism and accordingly promoted the motto "evolution as the control of development by ecology". The book provides some interesting details about Matsuda himself, including a few personal ones, and about the historical context in which his theories were developed. In particular, Reid's [304: 9] chapter is extremely valuable. For instance, it emphasizes the importance of Étienne Geoffroy St. Hilaire—the "father of Evo-Devo" for Panchen [289]—within this context because Étienne was one of the first to "deal empirically with epigenetic evolution because he realized that an accidental or environmentally imposed alteration in the development of an embryo would change the ultimate form of the mature organism." Therefore, the earlier the divergence in ontogeny, the more dramatic the change would be in the adult. Of course, "since Geoffroy was an anti-preformationist, and though that the environment acted directly on the embryo, he did not consider that the change might be a heritable mutation in the germ line of a parent" [304: 9].

According to Reid, "Geoffroy's evolutionism emerged from nature philosophy, which stressed 'unity of plan': the fundamental relatedness of all organisms; Geoffroy knew Lamarck and his laws but did not support Lamarckian progressionism; nor did he care for the particular hypothesis that evolution was caused by inheritance of characteristics which arose in response to a 'need' in the lifestyle of an organism." Instead, Étienne anticipated the Neo-Lamarckist emphasis on direct environmental causes of epigenetic change, and this is why Matsuda contended that he was a Neo-Lamarckist himself and that Lamarck's theories were much more complex and deep than suggested by most authors that have subsequently used him as a "straw man". Reid [304: 29] further noted that "some elements of so-called Neo-Lamarckism are Lamarckian, and the rest, in Buffon and Erasmus Darwin, are pre-Lamarckian; and a good grasp of the significance of direct environmental effects and epigenesis is found in the work of Geoffroy." Reid [304] explained that the major objection to Matsuda's Neo-Lamarckism is its dependence on, and the crucial role played by, the inheritance of acquired characteristics. According to Reid [304: 30], phenomena that Matsuda identified as a kind of inheritance of acquired characteristics, such as genocopying and genetic assimilation, are in fact examples of "the acquisition of heritable characteristics, since the requisite genes already existed." That is, for Matsuda "a true inheritance of acquired characteristics would require the kind of somatogenic mechanism that Darwin envisaged—pangenesis… no organisms since the ancient protobionts have been able to make brand-new genes", at least "never, as far as we know, by reconstruction to order."

Having said that, it should be noted that some epigenetic factors do seem to show a type of "Lamarckian effect", as pointed out by Schlichting and Pigliucci [327: 133] and discussed in the previous chapters of the present book. For instance, they reviewed works that examined the inheritance of dwarfism in rice (*Oryza sativa*) and showed that a single exposure to the chemical 5-azacytidine (azaC) immediately after germination was enough to produce dwarf plants that gave rise to a progeny that segregated into 35% dwarf and 65% tall types. Self-pollinization of the tall plants and of the dwarf plants resulted only in tall plants and dwarfs, respectively. Notably, analyses of the genomic DNA of these plants indicated that the azaC treatment had provoked a 16% reduction in DNA methylation relative to the parental plants; this pattern of methylation was maintained in the M1 and M2 generations, thus explaining the "Lamarckian effect". As explained in Chap. 5, although some authors continue to be highly reluctant to even mention Lamarck, it has become increasingly accepted that acquired traits, particularly behavioral and ecological ones (see Fig. 1.3) but probably also other types of epigenetic factors, can in fact be passed between generations.

Reid [304] and many authors of other chapters of the previously mentioned book, *Environment, Development and Evolution: Toward a Synthesis*, provided numerous examples of how epigenetic events influenced by the external environment, that affect organisms within the niche they constructed then became increasingly more independent of the original behavioral choices/patterns that led to the original construction of the niche. For instance, apart from many examples of symbiotic epigenetic effects [304], numerous empirical studies exist involving a predator that releases a diffusible molecule (e.g., a kairomone) that acts on the eggs of a prey species, which produce an phenotype not seen in the absence of the predator [171]. This is an emblematic illustration of how a single genotype can produce two different phenotypes in direct response to different environmental signals, i.e. of the existence of norms of reaction *sensu* Schmalhausen, a concept that is very similar to the current notion of phenotypic plasticity as noted by Hall [171]. One of Hall's favorite examples concerns the seasonal polymorphism of caterpillars of the geometrid moth *Nemoria arizonaria*. Although similar when they hatch on oak trees, the caterpillars that hatch in the spring mimic the catkins that are also on the oak trees in that season, whereas those that hatch in the summer (after the catkins fall) mimic twigs instead. The biotic signal for both morphs is the level of tannin in the trees. Therefore, it is easy to envisage how such adaptive phenotypes within a same population may be a first step toward incipient speciation ([171]: xix; for more details about the developmental mechanisms related to such polyphenisms, see [271] and references therein). By providing and discussing in detail the broader evolutionary implications of these and numerous other similar examples, the book *Environment, Development and Evolution: Toward a Synthesis* is highly informative and useful and greatly contributes to a change in mindset by stressing the importance of environmentally influenced epigenetic factors, which is key for ONCE and contrasts with the Neo-Darwinist gene-centered view of evolution.

However, as in Matsuda's works, I believe that the book by Hall et al. [171] also shows that even among most of the proponents of the Extended Evolutionary Synthesis, not enough emphasis is being given to the behavioral choices of the organisms themselves. This can be seen in Fig. 1.3 and is attested to by the fact that the word "behavior" does not appear in the title of that book, or in its main motto, or even among the hundreds of terms provided in its index. An illustrative example of the importance of behavioral shifts and their links to epigenetic factors, which also shows how, in that book, such shifts are often assumed to be merely secondary players in evolution, is provided by Reid [304: 21–22]. He wrote that "physiological stress for" the furry-footed Djungarian hamster "comes with extremely cold winters and short, dry summers; nursing mothers suffer a combination of dehydration and hyperthermia in the summer, due to lactation and the effective insulation that serves them so well in winter." However, he continues, "hormonal changes in both male and female hamsters *bring about* behavioral changes that diminish the mother's stress; the female forages, rehydrates, and cools off, while the male baby-sits; but if he had shown signs of being an old-fashioned absentee father, the mother would have aborted before coming to term." I used italics to emphasize how the behavioral changes are seen as "brought about" (synonym of "induced", "provoked", "the result of") by the hormonal changes, thus giving the appearance that organisms are merely passive players that can only be reactive to other, more important phenomena such as epigenetic factors, i.e. that organisms do not truly make their own behavioral choices.

In fact, when seen from a different angle this empirical case actually constitutes a powerful example of the importance of behaviors and the active evolutionary role of organisms. This is because it shows how the behavioral choices of the organisms lead to an interdependent cascade of events—including epigenetic phenomena related to hormonal changes, which in turn can be deeply related to genotypic effects [393]—that resulted in a specific successful evolutionary outcome. Importantly, this behavioral choice was not passive at all: it was one among many others that were possible within the behavioral/physiological plasticity of the organisms and the environmental context in which they lived. These choices included, for instance, the preference to stay in an extreme environment and to not migrate elsewhere, as many other animals do, and the subsequent preference of the females to rehydrate and cool off, and of the males to baby-sit. This is an illustrative case of the richness and diversity of possible behavioral choices in living organisms as well as their links with so many different forces, factors, and processes.

A similar example of the use of passive, even "negative" terms, used to refer to organisms and/or their behavioral choices in that same 2004 book is the statement, "spadefoot toad tadpoles, *Scaphiopus* spp., adopt a carnivorous form if they ingest shrimp; presence or absence of shrimp provides a reliable cue that induces tadpoles to transform into the alternative carnivore morphology; once tadpoles initiate the carnivore transformation, if they fail to ingest shrimp, they can revert to the omnivorous form" [339: 175]. First, the "presence or absence of shrimp" may be an appropriate sentence for a laboratory experiment, but in the wild the presence or absence of shrimp in a certain habitat/geographical place is of course related primarily to the initial behavioral choice of the shrimp to be, or to go, there or not. Second, the phrase "if they fail to ingest shrimp" gives the impression that, as usually contended by Neo-Darwinism, organisms are merely passive players that can only do two main things: survive if they do "what they are supposed to do" (and then reproduce) or die if they "fail to do so".

Standen et al.'s [346] recent experimental study is a particularly good example of a work stressing how behavioral shifts, particularly when linked to epigenetic factors, are crucial to drive major macroevolutionary changes such as those concerning the water–land transitions in vertebrates and the origin of tetrapods. They measured the developmental plasticity of anatomical and biomechanical responses in *Polypterus* fish experimentally reared on land, and showed remarkable correspondences between the environmentally induced phenotypes of terrestrialized

Polypterus and the ancient anatomical changes seen in stem tetrapods. In particular, terrestrialized *Polypterus* displayed less-variable walking behavior, planted their pectoral fins closer to the midline of their bodies, elevated their heads higher, and had less fin slip, thus allowing more effective vaulting of the more anterior (cranial) region of the body over the planted fin. The skeletal changes seen in the terrestrialized fish showed a manifest reduction in the external limits of the opercular cavity bordered by the supracleithrum and cleithrum bones, similar to that seen in tetrapodomorph fossil fishes such as *Eusthenopteron* (which are closely related to tetrapods), which seemingly facilitated greater flexibility between the pectoral girdle and the operculum. The authors hypothesized that the locomotory changes in the terrestrialized *Polypterus* likely affected the forces experienced by the skeleton, thus influencing skeletal growth and transforming skeletal shape, and that similar behavioral changes were present in stem tetrapods. That is, according to them, exposure to new or stressful habitats, such as a novel terrestrial environment, is a catalyst for releasing variation, and thus phenotypic plasticity can facilitate macroevolutionary change as a response to rapid and continuous environmental stresses. The authors postulated that a similar ontogenetic plasticity in Devonian sarcopterygian fish (the extant sister-group of tetrapods) in behavioral responses to terrestrial habitats might have facilitated the evolution of terrestrial features during the origin of tetrapods.

Lowe [238] recently summarized some insightful thoughts regarding epigenetics and its implications for evolutionary biology. As an illustrative example of a greater range of possible phenotypes—and, within one individual organism, of phenotypic heterogeneity—he referred to angiogenesis and the development of a complex vascular network in which the endothelium is in continuous interaction with the various tissue environments that it encounters. He pointed out that developmental plasticity depends on the ability of the organism not only to respond to the environment but also to use cues from it as signals to be interpreted by ontogenetic processes. These cues can therefore constitute informative causes of the outcomes of the developmental processes, and organisms can accordingly evolve processes that enable them to detect cues and react to habitat conditions to produce a functional outcome. In Lowe's words, "given plasticity, the range of relevant variation—be it internally or externally induced—that can produce functional organisms at a given end-point is therefore widened." He stressed that to understand the occurrence, perseverance and potential fate of differentiated cell types requires examination of a broader variety of causal factors than is usually performed in the literature. He challenged the conception of development as a "one-way process with a default way of unfolding", i.e. of "adultocentric" ontogenetic conceptions in which development is seen as a finalistic process of producing an adult from an egg.

Among other examples of the power of epigenetics published recently, one that I consider to be particularly insightful is that provided by Emmons-Bell et al. [123]. The example—which, I should note, should be subject of more scientific scrutiny, not only because of its astonishing implications but also due to the fact that it was not published in a top journal as one would perhaps expect for a paper with such remarkable   potential   repercussions—concerns   the   induction   of   different species-specific head anatomies in genetically wild-type *Girardia dorotocephala* flatworms. As the authors explained, it is commonly assumed that species-specific anatomical shapes are mostly "encoded" in the genome, although data from environmental epigenetics suggest that morphology is actually the combination of both genetic inheritance and environmental/life-history inputs. They thus focused their study on a specific type of biological regulation implicated in the epigenetic morphogenetic control: endogenous bioelectrical signaling. They note that spatio-temporal gradients of resting potential among all cell types—not merely excitable nerve and muscle—can influence such phenomena as cell proliferation, migration, shape, and apoptosis as well as large-scale morphogenesis and regulation of positional information, organ identity, size, and axial polarity.

Emmons-Bell et al. then refer to previous work that has implicated such voltage gradients in the regulation of antero-posterior polarity, appendage regeneration, craniofacial patterning, left–right asymmetry, eye development, and brain patterning. Next, they described their results, which revealed a striking stochastic phenotype, in which the regenerating heads of a genomically normal *G. dorotocephala* flatworms acquired—for 30 days—not only a head but also a brain morphology similar to that seen in other extant planarian species. Furthermore, anatomies corresponding to species that are most closely related to *G. dorotocephala*—which, however, split from it at least 100 million years ago—occur with a much higher frequency than morphologies corresponding to less closely related species. They thus concluded that form is both plastic and robust, that reinstatement of the "target" morphology can occur without trauma to the organism, and that such studies are thus crucial to illustrate the parameters and limits of anatomical variation.

A major question that needs to be addressed is this: if epigenetic factors, particularly those influenced by the external environment that are so crucial for ONCE, require the *a priori* existence of developmental plasticity to be flexible and be able to respond to different/changing internal/external factors, which phenomena lead to/allow the existence of such plasticity in the first place? A good succinct historical and integrative background on this issue was provided by Hall [168, 170), who stated that one of these phenomena is canalization. This term was coined by Waddington [372–380, 362, 363] for those features of developmental pathways that "buffer" physiological/metabolic systems against environmental and genetic perturbations. This buffering thus allows genetic/development variability to build up within the genotype without being manifested phenotypically. Similar buffering phenomena are sometimes designated "genetic homeostasis" or simply "developmental homeostasis". Waddington used the term "genetic assimilation" to link the hidden genetic variability produced by developmental canalization to adaptation and heritable phenotypic responses to environmental perturbations. Nijhout [271: 10] argued that the evolved insensitivity to variation can be seen as canalization (the capacity to return to a ontogenetic trajectory after a disturbance) or as robustness (the simple capacity to tolerate, or be insensitive to, environmental or genetic variation) and that, unlike physiological homeostasis, the mechanisms that confer ontogenetic robustness are not well understood.

Schlichting and Pigliucci [327] argued that canalization can be seen as a process leading to micro-environmental homeostasis and that canalization, plasticity, and homeostasis are not population-level phenomena being instead defined at the individual level. According to them, developmental instability is the opposite of homeostasis but has no necessary connection with plasticity in the sense that a given phenotype can be both plastic and stable as a reaction to macro-environmental stimuli and to micro-environmental noise, respectively. They defined two main types of genetic control of organismal plastic responses. The first —allelic sensitivity—concerns the response, which may be due to a change in the amount or activity of the gene product and vary among different alleles, by a specific gene locus to a shift in conditions. The second—regulatory control— concerns the existence of environment-dependent gene activity controlled by the action of a regulatory switch. The activity of the regulatory pathway can influence environment-specific expressions of one or various genes and thus lead to phenotypic-threshold responses or the emergence of polymorphism. According to this scenario, allelic sensitivity is a more passive type of plastic response that can be adaptive or not and that can lead to a diversity of pleiotropic phenotypic effects, whereas regulatory plasticity is more active, most likely adaptive, and mainly linked to the epistatic actions of plasticity genes. In addition, whereas regulatory control allows anticipatory responses to environmental change, allelic sensitivity does not.

Internal factors—such as those leading to phenomena such as canalization—are therefore crucial for ONCE because they both (1) allow/increase the power of Baldwin's organic selection by increasing developmental plasticity, and (2) constrain evolution by decreasing plasticity/variability. How can internal factors increase plasticity/variability? One of the most crucial ways is precisely through canalization, which can occur for multiple or individual phenotypes along a reaction norm, i.e. along all of the phenotypic states possible for a genotype within different environments [265]. That is, canalization leads to developmental directional phenotypic changes not by eliminating/constraining variability, per se, but by increasing it through robustness, i.e. by producing the same phenotype even if different genetic/epigenetic traits vary substantially, thus allowing the increase of so-called hidden developmental plasticity/variation.

Minelli [258: 59] provided an interesting discussion about the links between homeostasis and robustness by focusing on and reviewing cases in which desynchronization of development between different parts within a single organism still leads to a fully coordinated adult. For instance, when researchers applied a linear temperature gradient of approximately 10 °C for 5 h in either direction along the main body axis of the egg of the hymenopteran insect *Pimpla turionellae* (a parasitic wasp), this resulted in a dramatic ontogenetic desynchronization, lasting up to 9.3 h, between the egg poles. Within the same egg, up to seven mitotic waves were observed at the same time, and the cellularization process was exceptionally asynchronous. That is, ontogenetic processes that in normal development usually take place successively took place concurrently and vice versa. However, despite this strong disturbance, after the temperature gradient was switched off, the ontogenetic processes resumed their normal speed, thus pointing out that in such cases homeostasis leads to developmental robustness in the sense that even extreme and early ontogenetic desynchronization does not affect the segmental pattern of the resulting embryo.

Although apparently decreasing adult phenotypic plasticity by producing phenotypes that are more similar/canalized and distributed throughout a smaller morphospace, canalization can also increase the potential for plasticity by increasing the so-called "hidden plasticity/variation" that exists in the population but is not expressed in the adult phenotypes. Various examples of the latter phenomenon were recently reviewed by Morris [265]. One concerns populations living outside their tolerance limits: sea urchins exposed to low pH augmented transcript abundance for genes associated with biomineralization. This transcriptional plasticity restored the previous morphology despite the fact that the organisms were reacting to acidity levels that they never encountered in nature. As Morris pointed out, unexpressed plasticity may also be the result of the intrinsic ability an adjustment of organismal structure and behavior in response to new stimuli in a way that permits the functioning of the whole organism in novel environments, as exemplified in the 'two-legged goat effect' described in Chap. 1.

Unexpressed plasticity can also be caused by cryptic genetic variation, i.e. standing genetic variability that has no impact on phenotype or fitness under normal conditions but that can be manifested under stressful/new conditions. As summarized by Morris, cryptic genetic variation can be accumulated due to neutral mutations occurring in, for example, unexposed regions of a reaction norm, past selection in other habitats, or the action of phenotypic capacitors such as Hsp90 (see later text) that decrease the phenotypic effects of novel genetic inputs. Therefore, new conditions manifest these hidden variants, thus resulting in increased genetic variance for the feature. An important point for ONCE is that such phenomena further stress the significant role played by randomness in evolution. In this case, the random nature of this revealed plasticity may lead to some variants being more fit than others and a subset of the cryptic variation being selected, thus eventually leading to rapid adaptation as confirmed by empirical studies (see [265] for more details).

Morris also provides numerous other empirical examples linking developmental plasticity and epigenetics to behavioral and/or morphological changes. For instance, moss grown under zero-gravity conditions developed peculiar spiral morphologies; a killer whale at Marine Land learned to catch gulls by baiting them with its fish food and then passed on this acquired behavior to other whales; and the recent northern range dispersal of a butterfly species led to the production of a new wing coloration that was induced by temperature. In addition, unnaturally high carbon dioxide levels changed flowering time and carbon stores in several plant taxa, and invasive black rat populations produced a novel strategy for accessing pine seeds. Morris stressed that a combination of exploratory behavior and strong social learning can lead to behavioral innovation while also decreasing the risks of unnecessary exploration. He reviews empirical data showing that invading sparrows displayed greater innovation through their willingness to come close to and consume new types of foods than did long-established noninvasive sparrows of the same species, and that brain size is positively correlated with survival and invasive success, augmented innovation and tolerance of environmental uncertainty, and decreased mortality rates. In particular, an analysis of 446 introduction attempts in mammals revealed that successful colonization was significantly linked to brain size, even after taking into account body size, life history, mating strategy, and habitat generalism.

Hall [170: 148–151] noted that a major difference between the so-called "Baldwin effect" (Chaps. 1 , 2 and 3) and Waddington's genetic assimilation, recognized by Matsuda [245], is that Baldwin emphasized that organisms choose the environments that influence their development and that habitat or host selection thus occurs before the process of genetic fixation. Hall explained that this link between genotype and behavior was first demonstrated by Jean Piaget (famously known in particular for his work on clinical psychology and child development) in his examination of 80,000 specimens of the snail *Limnaea stagnalis*. Individuals with elongate shells were substituted by specimens with shorter shells, which were better adapted to the turbulent waters of many Switzerland lakes. Such shortened shells develop if snails are raised in turbulent water but are also formed when these snails are bred in still water. The proximate mechanisms are likely related to the contraction of the columnar muscles to allow developing snails to stick onto the substrate, with subsequent reduced growth of the shell. A behavioral response is also 'assimilated' in the sense that snails with shorter shells prefer shallow water. Therefore, according to Hall, both anatomical and behavioral features are "assimilated" in Piaget's study. He further pointed out that another renowned author who defended ideas similar to Baldwin's organic selection was Schmalhausen, who saw adaptation to a novel habitat as the first step toward destabilizing ontogeny and allowing previously hidden genetic/ontogenetic plasticity to be manifested phenotypically.

As reviewed by Gilbert [148: 236–237], Waddington's idea that genetic assimilation is linked to canalization and occurs in nature remained controversial until 1998, when Rutherford and Lindquist revealed a possible molecular mechanism for it. The anomalies they observed when Hsp83 was mutated, or the Hsp90 protein produced by this gene was inactivated, did not show simple Mendelian inheritance being instead the result of interactions of numerous gene products. Selective breeding of the anomalous flies that they studied led, over a few generations, to populations in which 80–90% of the descendants had the mutant phenotypes in which the mutants did not retain the Hsp83 mutation. That is, once the mutation in Hsp83 allowed the cryptic mutants to become expressed, selective matings could keep the abnormal phenotype even in the absence of abnormal Hsp90.

Therefore, Hsp90 appears to be a critical feature for canalization and the related increase of hidden variation in the sense that it allows mutations to accumulate but prevents them from being manifested until the environment changes. As Gilbert pointed out, temporary decreases in Hsp90 would expose preexisting genetic connections that generate anatomical variations that would likely be deleterious. However, some of these variations might be selected in the novel habitat; therefore, continued selection would enable the fixation of adaptive physiological responses to the habitat. Badyaev [22] stated that empirical studies have suggested that epigenetic integration often precedes genetic integration of new features. Therefore, novel features originating in the interactions between an organism with its environment can remain epigenetically integrated for an expanded time before acquiring greater genetic integration by way of natural selection and becoming incorporated into the lineage's ontogenetic stock. Badyaev explained that these results were supported by computer simulations showing a propensity for emergent processes to gradually develop hierarchical controls. As ontogenetic complexity of local adaptations increases, this propensity results in greater genetic determination of upstream ontogenetic stages and stronger epigenetic and environmental influences on later stages.

Box—Definitions: Baldwin effect, Waddington's genetic assimilation, West-Eberhard's Genetic Accommodation, and ONCE

A lucid comparison between the Baldwin effect and Waddington's genetic assimilation in the framework of the current Extended Evolutionary Synthesis (Fig. 1.3) was made by Crispo [64]. As she noted, Baldwin often referred to "accommodation" as non-heritable phenotypic changes that occur in response to the environment and enhance the organism's survival in the particular habitat where the phenotype is induced. Phenotypic accommodation *sensu* West-Eberhard broadly corresponds to Baldwin's accommodation, except that it incorporates responses to both genetic novelties and environmentally induced changes, whereas Baldwin's accommodation referred only to the latter. West-Eberhard's phenotypic accommodation is defined as "adaptive mutual adjustment, without genetic change, among variable aspects of the phenotype, following a novel or unusual input—genetic or environmental—during development." She defined genetic accommodation as a different phenomenon, linked to adaptive genetic changes.

Crispo noted that a further difference between Baldwin's and West-Eberhard's concepts is that Baldwin's may also include labile, i.e. non-developmental, plastic shifts. In addition, Baldwin defended the notion that heritable ("coincident") variations can occur in the same direction as the plastic response and that phenotypes that are initially environmentally induced can thus be selected upon and inherited. Crispo argued that this phenomenon, which is supported by various empirical examples reviewed in her article, is currently considered a type of West-Eberhard's "genetic accommodation". She noted that the "two-legged goat effect" mentioned in the beginning of the present book is an example of West-Eberhard's phenotypic accommodation because the goat was able to adopt bipedal locomotion not only through a behavioral shift but also by way of plastic ontogenetic morphological changes such as the enlargement of the hindlimbs and changes in the vertebral spine and pelvis.

Table 8.1 shows Crispo's summary of the major differences between the Baldwin effect, Waddington's genetic assimilation, and West-Eberhard's genetic accommodation, and Fig. 8.1 further illustrates the differences between the former two phenomena. Crispo explains that Waddington proposed, as noted previously, that development would evolve to become canalized against environmental perturbations by way of selection acting on the developmental system, i.e. genetic assimilation. In this sense, it is interesting to see how Waddington's idea that, during development and also during evolution, features/organisms would be buffered against external perturbations is in line with modern systems biology thinking, which argues that there is an evolutionary trend toward increased homoeostasis and thus decreased dependence on the external environment/perturbations (see [50]). Crispo then goes on to note that although Waddington's idea may seem conceptually comparable with the Baldwin effect, there are essential differences between the two. Waddington defended the idea of an "epigenetic landscape" where the rolling of a ball across a landscape symbolized ontogeny and where environmental perturbations could push the ball from one developmental pathway to another. Genetic assimilation thus acted as an evolutionary process to amplify the landscape ridges so that, over time, increasingly greater perturbations would be needed to shift the ball from one ontogenetic trajectory to another. In other words, according to Waddington the environment induces phenotypes that are adaptive, and then selection on the ontogenetic system acts to decrease the reactions to the environment—i.e. to reduce plasticity—so that the induced phenotype becomes "inherited", i.e. canalized, after various generations of exposure to the environmental stimulus. As Crispo noted, Waddington's idea was therefore similar to Schmalhausen's [328] definition of "stabilizing selection" as selection acting against both mutations and environmentally induced shifts that push the

Table 8.1 Mechanisms of evolutionary change

	Basis for trait inducing the evolutionary response	Increase or decrease in the level of plasticity	Mean phenotypic value in the inducing environment
Baldwin effect	Environmental	Neither or increase (can also include a decrease in plasticity, although Baldwin focused less on such cases)	Changes
Waddington's genetic assimilation	Environmental	Decrease	Stays the same
West-Eberhard's genetic accommodation	Environmental or genetic	Neither or either	Changes or stays the same (inducing environment could include either the extrinsic environment or the genetic environment)

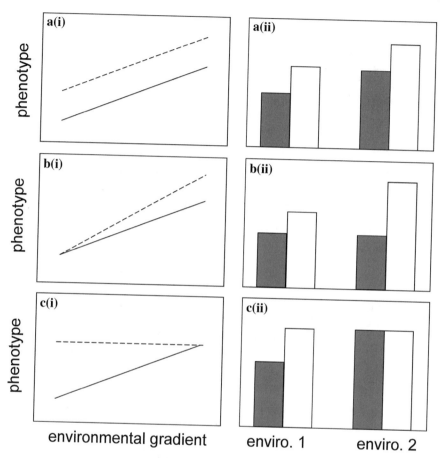

Fig. 8.1 Relationship between phenotype and environment, i.e., plasticity before (*solid lines, gray bars*) and after (*dashed lines, white bars*) genetic accommodation according to, and modified from Crispo [64]. **a, b** Baldwin effect. **c** Waddington's genetic assimilation. (i) Reaction norms for continuous phenotypic variation along an environmental gradient and (ii) polyphenisms in discrete environments. The height of the bars represents either the magnitude of the induced phenotype or the frequency of the polyphenism in the population

phenotype from the local optima. The main difference is that Waddington proposed that environmental perturbations would initially be adaptive, whereas Schmalhausen defended that they would be chiefly, but not always, maladaptive.

After summarizing and clarifying the theoretical differences between Waddington's genetic assimilation and the Baldwin effect, Crispo recognized that some of Waddington's experiments actually illustrate how both phenomena can be confused in practice. For instance, Waddington [381] raised larvae of three strains of the fly *Drosophila melanogaster* in a salt medium that lead to at least 60% mortality of the flies. He raised the flies for 21 generations and augmented the salt concentration each generation to maintain a constant mortality rate. After these 21 generations, he raised larvae from the selected lines and from the original stock flies on media with differing levels of salt. Waddington noted that even though the unselected stock tended to have higher survival at lower salt concentrations, the selected lines had higher survival at higher salt concentrations. In all cases, the survivors in the selected lines had greater areas of anal papillae, which are involved in osmoregulation, and anal papillae area increased with increasing salt concentration. He then proposed that the anal papillae increased area in the selected lines provided support for his notion of genetic assimilation. However, according to Crispo, this is a contradiction of Waddington's own definition of genetic assimilation as a process leading to canalization, i.e. to augmented genetic control of a phenotype. This is because in this case the phenotype that was expressed in higher salt concentrations was not canalized in the selected lines as shown by the reaction norm. Instead, it appears that the salt-selected larvae had an even steeper slope of this norm, thus suggesting that plasticity had increased. According to her, it is possible that the most plastic specimens were the ones most capable of survival, and thus Waddington might have inadvertently selected for plasticity, i.e., his case study does not categorically validate his notion of genetic assimilation but rather appears to demonstrate the Baldwin effect (see Fig. 8.1 and Table 8.1).

Although I particularly admire the paper by Crispo [64], from which Table 8.1 is based, for its lucidity—and accordingly decided to pay special attention to it in this chapter and the previous box in particular—I should note that I do not completely agree with a few of Crispo's points. For instance, she wrote: "although Waddington did not completely denounce Neo-Darwinism, he highlighted the deficiencies in the theory, namely that random mutation is unlikely to produce all the variants selected upon in nature, and that the importance of environmental induction in evolution should not be ignored" [64: 2469–2479]. She further noted that "Waddington criticized Baldwin's theory, stating that the accumulation of genetic mutations influencing an induced trait is unlikely; yet, both theories rest on the assumption that natural selection acts upon favorable mutations, and are thus both fully compatible with, and in fact rely upon, Neo-Darwinian evolution."

I do not fully agree with these latter statements because Neo-Darwinism is not simply about selection acting upon favorable mutations; if not it would apply to everything and therefore explain nothing at all. Neo-Darwnism argues specifically (1) that random mutations occur first and lead to random phenotypic changes and (2) that only subsequently are the new phenotypic features that happen, by accident, to be favorable in a certain habitat, or ecological/behavioral context, selected (see Table 2.1 of Chap. 2). As explained in Chap. 5, Neo-Darwinism thus predicts that form would normally change before function/behavior and that natural selection is the key primary player in evolution. This clearly contrasts with Baldwin's organic selection, and for that matter with ONCE (Fig. 1.2), in which there is first an organismal behavioral response to changes in the environment (or simply a behavioral shift in the same environment) and subsequently the random mutations that lead to phenotypic features that are favorable within the new behavioral/ecological context are selected by natural selection (Table 1; Fig. 1.2). Thus, both Baldwin's organic selection and ONCE predict that function/behavior would normally—but not always, e.g. 'hopeful monsters' could be an exception as noted above and as will be explained in more detail below—change before form. In addition, both organic selection and ONCE envisage the organisms themselves, in particular their behavioral choices and persistence, as the main drivers of evolution with natural selection being a secondary, but still crucial, evolutionary player.

As noted by Balon [27: 42–43], authors often refer to Waddington's notions of homeostasis, canalization, and genetic assimilation but not as often to his concept of homeorhesis, which is crucial for ONCE: "in contrast to homeostasis as a process keeping something at a stable or stationary state, Waddington proposed for living systems the term homeorhesis, meaning stabilized flow… the stabilization of a progressive system acts to ensure that the system goes on altering in the same sort of way that it has been altering in the past… therefore we may define, for our purposes, any steady state as homeostasis and any stabilized state as homeorhesis." That is, homeorhesis can in theory lead to major, organized changes by stabilizing them. According to Balon, "during a stabilized state… the homeorhetic processes of the system 'resist' destabilization for as long as possible, enabling structures to be completed and functions to progress without interfering with stabilized life activities." However, when a system is destabilized, "a switch is rapidly made via a far from stable threshold into the next stabilized state of ontogeny… the system will assume a new steady state on the crossing of the threshold, and the resulting phenotypic transformations will then depend on the reaction norms of the system at this point, as well as on the secondary reactions of associated systems." Using his own words, Müller—who defends a physicalist view of evolution as noted previously —"emphasizes that thresholds are an inherent property of developing systems, able to trigger discontinuities in morphogenesis which can automatically result in the generation of a new structure; novelty can thus arise as a side effect of evolutionary changes between two self-organized and maintained states."

In this sense, Waddington's homeorhesis can play a similar role in directing developmental changes as behavioral persistence associated with natural selection play in directing macroevolution. Moreover, the notion of homeorhesis may be useful to explain the occurrence of saltatory ontogeny and saltatory evolution *sensu* Balon, being somewhat related to the notion of "facilitated variation" *sensu* Gerhart and Kirschner [144, 145]. Gerhart and Kirschner [145: 8582] proposed that "the number and kinds of regulatory genetic changes needed for viable phenotypic variation are determined by the properties of the developmental and physiological processes in which core components serve, in particular by the processes' modularity, robustness, adaptability, capacity to engage in weak regulatory linkage, and exploratory behavior." Therefore, "these properties reduce the number of regulatory changes needed to generate viable selectable phenotypic variation, increase the variety of regulatory targets, reduce the lethality of genetic change, and increase the amount of genetic variation retained by a population; by such reductions and increases, the conserved core processes facilitate the generation of phenotypic variation, which selection thereafter converts to evolutionary and genetic change in the population."

That is, within this view it would not be unexpected that Goldschmidt's "hopeful monsters" could occur—and eventually survive and in some cases potentially thrive—in evolution. For instance, at first it might seem very difficult to explain how a full extra appendage might appear, with not only an internal skeleton but also muscles, nerves, blood vessels, skin, or other type of connected/enveloping tissues, and so on, as it sometimes does in human congenital malformations. There are various reports of fetuses and neonates with three forelimbs, for instance, and although the extra limb is usually not functional, it does have bones, muscles, nerves, and other tissues that have anatomies and patterns of connection that are, at least partially, comparable with those of normal limbs, as I can attest after dissecting several extra limbs in humans and other animals such as salamanders. By taking into account Waddington's notion of homeorhesis and Gerhart and Kirschner's notion of facilitated variation—which are in a way very similar to Goldschmidt's notion of "regulation"—it is actually not so difficult to explain these phenomena. This is because if a limb bud occurs/is experimentally placed in an abnormal position, it may still lead to the formation of a limb with an internal skeleton, which is developmentally associated with the formation of muscles (see [111]) that are in turn linked to the formation of nerves and blood vessels, and so on. This was precisely the type of argument made by Balon [27], who is a strong proponent of the notion of saltatory evolution.

Another point that is often not mentioned but that becomes very clear when one reads Goldschmidt's original works is that he defended that heterochrony would be often a crucial event leading to "hopeful monsters". This idea that heterochrony can lead to major macroevolutionary changes is in fact now commonly accepted in evolutionary biology. Of course, as stressed again and again by Goldschmidt, in the early stages of their evolution, those "hopeful monsters" would need to benefit from a partially relaxed selection, somehow, and this also fits with the idea of non-optimal, non-struggling evolution that is defended in the present book and is one of the tenets of ONCE.

A new sub-field of Evo-Devo, Evolutionary Teratology *sensu* Guinard [160] and Evolutionary Developmental Pathology (Evo-Devo-Path) *sensu* Diogo et al. [106] —Evo-Devo-P'Anth is a subfield of that field that is directly related to human evolution/anthropology, thence the "Anth" after the "P" for pathology—is based on the notion that current and extinct wildtypes of certain taxa are essentially examples of hopeful monsters. Guinard [160] noted that Goldschmidt [152] introduced the concept of developmental macromutations—i.e. mutations of chief developmental genes that can produce substantial phenotypic effects—to explain macroevolution. Although the majority of such mutations would be disastrous (a "non-hopeful monsters"), there may be one macromutation leading to subsequent adaptations of an organism to a new way of life, i.e. to a "hopeful monster". Therefore, in Guinard's view macroevolution would occur with the rare success of "hopeful monsters" rather than by an accumulation of small changes in populations. As explained by Guinard, this notion is, however, often misinterpreted—as are many other ideas of Goldschmidt, who is so often used as a "straw-man" together with Lamarck and Haeckel in evolutionary biology (see Chaps. 5 and 7)—as the achievement of "perfection" in a jump (extreme saltationism).

Guinard argued that the relatively short forelimbs of dinosaurs, such as the Tyrannosauridae (which includes the famous "T-Rex"), display features that are strikingly similar to features associated with well-known human pathologies (e.g. micromelia) (Figs. 8.2, 8.3, and 8.4). Such examples moreover further support Alberch's Logic of Monsters by stressing that pathological features of one taxon (e.g. humans) are often seen in the normal phenotype of another taxon (e.g. dinosaurs), thus leading to a resurgence of the ideas defended by Isidore Geoffroy Saint-Hilaire (Étienne's son) more than 150 years ago, who stated: "monstrosity is no longer a blind disorder but another order, also regular and subject to laws" [160: 21]. A brief historical account about the influence of authors—such as Goethe and both Isidore and Étienne Geoffroy Saint-Hilaire for the development of "Evolutionary Teratology" *sensu* Guinard—is given in the box below. This brief account is mainly based on Diogo et al.'s [113] recent paper on the links between the ideas of these three authors and Waddington's homeorhesis, Alberch's logic of monsters, and Goldschmidt hopeful monsters in particular and between pathology, teratology and macroevolution in general.

Box—History: Goethe, Étienne, Isidore, Waddington, Goldschmidt, 'Monstrosity', Teratology, and Evolution

Although the terms Evolutionary Developmental Pathology (Evo-Devo-Path) and Evolutionary Teratology were only formally defined in two papers published in 2016 (see previous text), various key broader issues addressed in those papers have been discussed by previous authors since a long time ago. In particular, of the three main components of Evo-Devo-Path, pathology (particularly human congenital malformations) has interested both erudite and lay people since thousands of years ago (including Aristotle: [223]). One of the key persons that influenced many of the researchers that have attempted to discuss both the reasons behind and the broader implications of human congenital malformations—based on the systematic compilation of empirical data—after the so-called scientific revolution was Goethe (lived in 1749–1832). This is because, as noted previously, Goethe was one of the earlier more prominent defenders of internalism, which had a great impact in the Romantic German school (e.g. Oken) and Naturphilosophie (e.g. von Baer). As also explained previously,Bateson [34], who was influenced by Goethe's ideas, compiled an impressive number of studies about animal morphology, human development, variations, and defects and defended ideas that are now becoming mainstream in Evo-Devo.

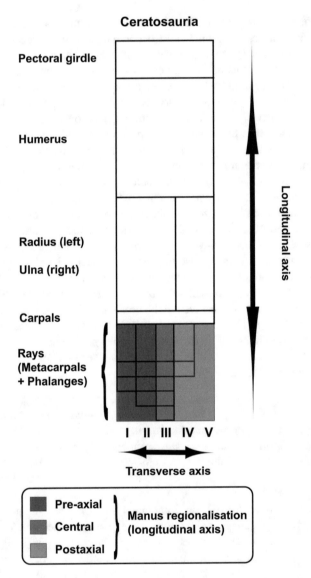

Fig. 8.2 Mapping of forelimb of the dinosaur clade Ceratosauria with various parameters used for the nomenclature of teratological features in humans. Modified from Guinard [160] and Diogo et al. [113]. Note that the central and postaxial regions may not be the same because of the number of digits (as with the mapping of Tyrannosauroidea—excluding Proceratosauridae—where the third and last digit is considered as postaxial)

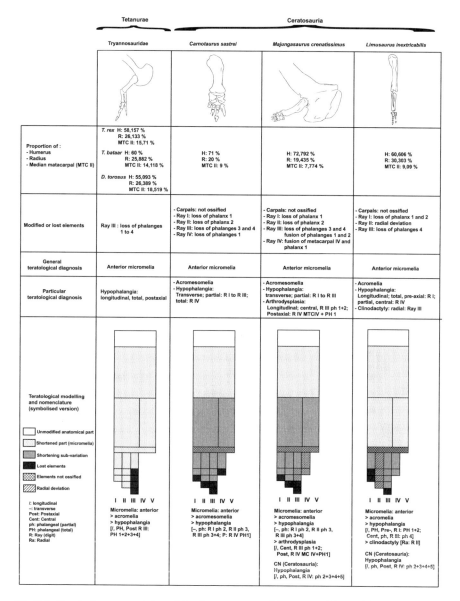

Fig. 8.3 Forelimbs representative of the dinosaur groups Tyrannosauridae and Ceratosauria with severe shortenings of the forelimbs along with anatomical anomalies and a scheme of teratological identification and nomenclature. Modified from Guinard [160] and Diogo et al. [113]. *Gorgosaurus libratus* and *Albertosaurus sarcophagus* are also related to the modelling and nomenclature of Tyrannosauridae. Not represented here is *Aucasaurus garridoi* (Carnotaurinae), which exhibits other manual variations (see [160] and Diogo et al. [113]) for more details. CN is for complementary nomenclature for concerned taxa, e.g. digit no. 4 of Ceratosauria is not complete: it is a previous teratological feature persisting along the lineage that must therefore be repeated for each taxon

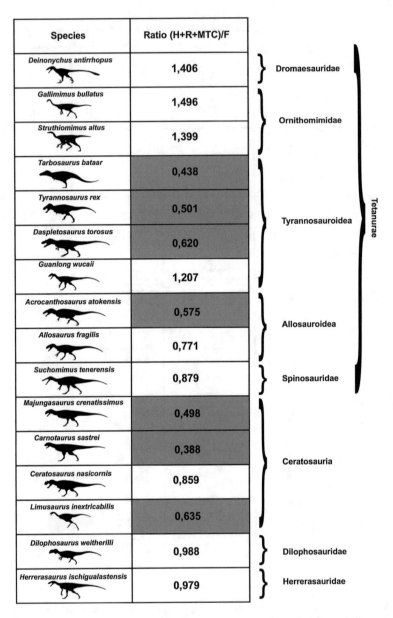

Fig. 8.4 Ratio of comparison between forelimb-length proportions (humerus, radius, median metacarpal) in relation to the femur length among theropod dinosaurs. Modified from Guinard [160] and Diogo et al. [113]. The shaded cells indicate species with an anterior micromelia (ratio ≤ 0.66)

As also noted previously, Étienne Geoffroy Saint-Hilaire (lived in 1772–1844), considered by some authors to be the "Father of Evo-Devo", was also influenced by Goethe's view of nature: Étienne's evolutionism "emerged from nature philosophy, which stressed 'unity of plan'... he had realized that an accidental or environmentally imposed alteration in the development of an embryo would change the ultimate form of the mature organism; so, the earlier deviation of embryonic development, the more magnified the change would be in the adult" [304: 9]. These links between development and deviation of form, which were the subject of experimental studies by Étienne [116, 315], were further explored by Étienne's son, Isidore Geoffroy Saint-Hilaire (lived in 1805–1861). The term "teratology"—the science of anomalies in the organization of living beings—was actually introduced by Isidore [143]. A current definition of teratology underlines a "discipline devoted to the study of congenital morphological anomalies, their causes and teratogenesis" (Dictionnaire Médical de l'Académie de Médecine, version 2016-1).

For Étienne, "monstrosity" was at a high, special level among anatomical anomalies. His intention was to consider "monsters" without prejudices and study them through the identifiable and classifiable facts. Thus, it was necessary to also think beyond the deviation from the original shape [13]: Étienne wrote "all this false position comes, I think, of what the eyes saw with so much evidence has not been perceived by the mind; one was accustomed to deal with beings from the point of view of their distribution by species, and monsters were seen in the spirit of this routine and not in the way they stroke our senses" [142: 108–109]. By means of Étienne's theory of the unity of organic composition—or the continuity of organisation between beings—"monstrosity" was discussed within an overall framework of the organization plan through rules, including the "principle of connections" or the "balancement of organs" [13]. That is, for him order exists in the apparent disorder, which leads to establish a classification grouping individuals under the heading of common deformities.

The study of teratology has thus emerged within a comprehensive theoretical framework of the organization of life, with the ability of features to be passed from one species to another through transformation. Many of Étienne's ideas are still present in current Evolutionary Teratology (*sensu* [160]). Yet, as stressed by Guinard, something was still missing in Étienne's conceptualization of the unity of organic composition: evolution as it was later defined by authors such as Darwin and Wallace. Evolutionary Teratology *sensu* Guinard is precisely an attempt to bridge this conceptual and theoretical gap, as explained by Guinard [160] and by Diogo et al. [113].

Interestingly, Guinard [160] considers that the term "hopeful monsters" should actually be abandoned because Evolutionary Teratology should instead focus on two important points: (1) the occurrence of anatomical abnormalities (in all degrees of expression, even extreme variations); and (2) the evolutionary success of concerned organisms (fossils and evolving lineages). He provides the example of Testudines (turtles), those puzzling vertebrates with limbs originating inside the rib cage that are often called "hopeful monsters". For him, they cannot be considered "monsters" nor "hopeful" in the sense that of their fossil record extending > 200 million years and their current broad distribution around the globe. He explains that he prefers to qualify turtles as animals exhibiting a viable teratology that was not particularly challenged by circumstances: their peculiar features must be considered modifications, peculiar, or outstanding, but above all as variations; a prosperous teratology thus must be considered as a form of success whether or not it is considered an "adaptation". In fact, turtles would also not fit the notion of "hopeful monsters" in the sense that the more fossils that become known, the more it is recognized that the transitions toward the forms displayed by extant turtles were actually gradual ones [209]. Therefore, it is likely that at least in some cases what may seem to be evidence of saltational evolution is just an "illusion" due to a lack of an appropriate fossil record, although—as will be explained later in the text— there are also many well studied cases for which there is a broad fossil record and that do seem to be the result of evolutionary saltations.

Schlichting and Pigliucci's [327] book provides another clear example of the renewed interest in the work of Goldschmidt and recognition that his view of macroevolution was much more complex and comprehensive than the caricature that Neo-Darwinists have tried to create. Schlichting and Pigliucci [327: 34] state that Goldschmidt actually "pointed out plentiful experimental evidence for the dramatic phenotypic effects of macromutations, for example the contrast of the normal diploid tobacco plant with several trisomic lines originates by chromosomal mutation; the range of morphologies and architectures was staggering, and all the plants were healthy and capable of surviving under natural conditions" (see Fig. 8.5). I would add that the criticism of Goldschmidt's "hopeful monsters" by Neo-Darwinists was facilitated by a biased, anthropogenic view of evolution. In humans, any "macromutation"—such as a chromosomal duplication leading to trisomy, except in the case of trisomy 21 associated with Down Syndrome—tends to lead to the death of the individuals well before they reach sexual maturity and thus before they are able to leave descendants ([342]; see previous text). However, in many other animals, and particularly in other organisms such as plants, similar macromutations very often do not have the same dramatic effect as they have in humans. Instead, they may produce healthy and viable organisms as shown in Fig. 8.5 and evidenced by the huge chromosomal diversity seen in not only different taxa but also within the members of a single taxon (see, e.g., Levin's 2002 book *The Role of Chromosomal Change in Plant Evolution*, and references therein).

An example of wildtypes of current taxa that could be seen as "hopeful monsters" concerns a group of marsupials in which syndactyly (fusion) of digits 2 and 3

Fig. 8.5 Normal diploid tobacco plant (*Nicotiana tabacum*) (*top left*) plus five trisomic variants that are healthy and capable of surviving under natural conditions. Modified from Goldschmidt [152]

is the normal phenotype. The group was even called "Syndactyla" for some time, but today it is generally considered to be non-monophyletic [391]. As these latter authors note, such cases of syndactyly are probably homoplastic in marsupials and may well be due to non-adaptive features linked to the integration of digit elements through ontogenetic constraints. That is, syndactyly would merely represent the loss of a single digit rather than the incapacitation of both digits as is the case with the soft-tissue syndactyly of the human hand. Thus, in functional syndactylous digits the net loss of a single digit may have had a mild impact compared with the loss of flexion ability that occurs in syndactylous human hands. According to these authors, this may have thus contributed to the proliferation of marsupial syndactyly within the ancestral populations, i.e. it would be in essence a neutral feature in this respect, and not as detrimental as in human syndactyly.

As they also note, digital anatomies that are in at least some ways similar to marsupial syndactyly were displayed by at least some of the earliest tetrapods, such as *Acanthostega*, and as such theses similarities have been attributed to identical patterns of morphogen expression in the digits. Moreover, identical digital morphologies have also been produced by way of biochemical treatment of developing autopodia (hands/feet) in mice, chickens, and frogs. Such experiments basically used alteration of the expression or concentration of morphogens—mostly Bmp family transcription factors— and, interestingly, many of the studied outcomes displayed retention of the inter-digital membrane as occurs in syndactylous marsupials. Weisbecker and Nilsson argued that there was likely a rapid origin of fully integrated/functional syndactylous digits in marsupials, which would also explain the lack of intermediates of syndactylous feet in both extant and fossil marsupials as well as the heterochrony observed in their study. In fact, they noted that similar scenarios with a localized change in Bmp-expression patterns have been suggested for other cases, such as the origin of bat wings, and argued that such scenarios would clarify the conservative size relationship of digits across bat species and the lack of fossil intermediates of bat-wing evolution.

Another fascinating case concerns chameleons, which are slowly becoming a model organism in both Evo-Devo and medicine (see Fig. 8.6). As noted by Diaz and Trainor [79], these reptiles have limbs that are seemingly adapted for an arboreal lifestyle including peculiar features such as a midline autopodial cleft (ectrodactyly) and two opposable syndactylous bundles of digits (clusters in the hands and feet that retain interdigital tissue) that are highly mobile (zygodactyly). Developmental studies of chameleons are particularly interesting because they may elucidate possible mechanisms that have led to the occurrence of such "hopeful monstrosities". For instance, cleft formation in hands/feet is currently assumed to be linked to a failure to keep the integrity of the apical ectodermal ridge, principally along the distal midline. Loss of this ridge inhibits distal outgrowth of the limb and leads to the loss or splitting of the digital rays.

In a study by Diaz and Trainor [79], Fgf8 was observed in this ridge, whereas Shh was present in the zone of polarizing activity from the beginning of limb-bud outgrowth by way of termination of distal-limb outgrowth and morphogenesis. That study thus contradicted the idea that autopodial ectrodactyly and clefting occurs due to a loss of integrity of the apical ectodermal ridge because in chameleons this ridge remains a robust ectodermal thickening and expresses Fgf8 during proximal mesenchymal clefting. Interdigital cell death, an ancestral tetrapod condition, was mainly seen in chameleon autopodia between the digit pairs undergoing clefting, whereas adjacent digits retained interdigital mesenchyme (syndactyly). Interestingly, inhibition of Bmp signaling before the beginning of interdigital cell death prevented cleft formation. However, when performed after cleft formation had started, it had no effect on interdigital cell death. Cell death between syndactylous digits was present in chameleons but at

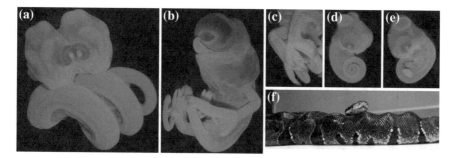

Fig. 8.6 Anomalies present in squamate reptiles that have also been identified in humans, thus underscoring the appropriateness of natural and malformed reptiles as models for human diseases (mutant embryos were harvested during normal laboratory activities not induced through mutagenesis). Modified from Diogo et al. [113]. **a** Dicephalic African house snake (*Boaedon fuliginosus*), generally due to excessive hedgehog signaling along the anterior embryonic midline. **b** Cyclopic (Holoprosencephalic) veiled chameleon (*Chamaeleo calyptratus*) whose deficiency is generally due to an absence of hedgehog signaling along the anterior midline. **c** Unilateral polydactyly and an ectopic limb branching from the knee of the same cyclopic individual. **d** Amelia of forelimbs and hindlimbs (phenotypically very snake-like). **e** Amelia of the forelimbs only. **f** A putative Ball python (*Python regius*) model displaying an ALS (amyotrophic lateral sclerosis, Charcot's disease, Lou Gehrig's disease) phenotype

significantly inferior levels compared with the clefting mesenchyme. The syndactyly seen in chameleons may thus differ from that observed in duck feet and bat wings due to the absence of Grem1, a Bmp antagonist, in interdigital tissue.

Strong theoretical and empirical support for the importance of "hopeful monsters" for evolution, and in particular for macroevolution and the origin of new bodyplans, is given in Raff's [302: 189] book, *The Shape of Life*. One of the many examples he presents concerns the importance of homeotic transformations in the evolution of insect bodyplans. Both the clade Diptera (Fig. 8.7), which includes houseflies, and the clade Strepsiptera (Fig. 8.8), which includes insects commonly known as twisted-wing parasites, have individuals with a single pair of wings and a pair of halteres. However, in dipterans the wings are on the second thoracic segment and the halteres on the third, whereas in strepsipterans the main wings are on the third and the second segment has shortened forewings that, according to Raff, correspond to/derive from dipteran-like halteres. As noted by him, it is thus possible that the Ubx gene that is expressed in the *Drosophila* third thoracic segment is expressed in the second thoracic segment of strepsipterans. Such a shift might not be detrimental, i.e. having wings and halteres in the second and third versus third and second segments may be a functionally neutral change, or even be advantageous depending on the behavioral and ecological context in which the last common ancestors of strepsipterans lived. Thus, a potential homeotic transformation that might have occurred by chance could have led to a different form and ultimately to the existence of a new higher clade. Raff provides numerous similar examples, particularly among insects, in his book.

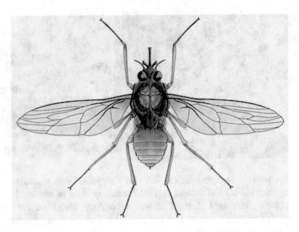

Fig. 8.7 Dorsal view of a fly (order Diptera, class Insecta). Drawing by Andrew Howells. Modified from http://australianmuseum.net.au/flies-and-mosquitoes-order-diptera, public domain

Fig. 8.8 Dorsal view of a male of *Xenos vesparum* (order Strepsiptera, class Insecta). Modified from [206]

The crucial importance of homeotic changes for macroevolution is particularly well illustrated by a beautiful, simple series of fascinatingly diverse silhouettes made by a non-scientist (Fig. 8.9). Concerning this subject, Raff [302] argues that the concept of a developmental bodyplan is related to the communality of onto-genetic processes leading to an adult structure, which is probably related to a conservative phylotypic stage in the sense that the internal constraints allow no other way to build a certain bodyplan. Then, within this communality, single-switch gene determination or ontogenetic modules can occur at various developmental stages. For instance, simple signals and genes related to binary choices are critical for segmental identification and the demarcation of segment-specific modular

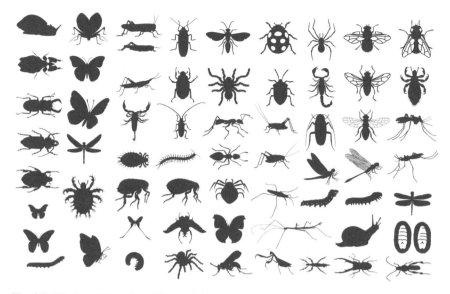

Fig. 8.9 Display of diversity of forms within invertebrates, particularly within insects. Modified from http://findpik.com/img/8717249 (Vector EPS, public domain)

structures in the insect body axis and vertebrate brain and in the specification of individual organ primordia. For example, ectopic expression of the nanos gene product in the anterior (cranial) end of a *Drosophila* embryo leads to the development of a second abdomen. According to Raff, although modules may have complex internal gene-expression states, expression of only a single gene may be necessary to direct them to particular attractors. The product of such a selector gene commits cells expressing this gene to a state defining the formation of a specific module, thus leading to major developmental and evolutionary consequences. That is, substantial evolutionary changes can occur by way of small changes in existing gene networks without reinventing the basic components, cell types, tissues, and organs of the body of organisms.

Another example provided by Raff [302: 412] concerns the expression of the gene wingless in a segmentally iterated manner in embryos of the genus *Tribolium* (which includes, e.g., flour beetles). The fact that the wingless-expressing cells lie in the same relationship to the engrailed-expressing cells, as seen in the fly *Drosophila*, suggest an old origin for, and conserved pattern of, such a mechanism of gene expression. Raff pointed out that—as explained previously and shown in Fig. 7.2—in plesiomorphic insects of the Paleozoic (era from 541 to 252 million years ago) the leg and wing primordia might have arisen in all thoracic and abdominal segments, that the wing primordia migrated dorsally as they do in dragonflies, and that the adults would thus have had legs and wings on all body segments. Then, as Paleozoic insects evolved further, there was likely an evolutionary trend toward the prevalence of thoracic legs and wings, the suppression of abdominal legs, and wings being controlled by the homeotic genes of the *Antennapedia* and *Bithorax* complexes (see previous notes on heterotopy).

The occurrence of such macroevolutionary major modifications, and of "hopeful monsters", thus appears to be an exception to the more common occurrence of "function/behavior changes before form" predicted by ONCE, as will be further discussed in this chapter. However, it is possible that, even in these cases related to major saltational transformations, some of these modifications also occurred after behavioral shifts and were subsequently selected in view of the new behavioral context and niche constructed by those shifts. This is illustrated by one of the examples provided by Raff (2016: 360). Suppose that greater body flexibility, for instance linked to a new behavior in which such flexibility would be advantageous, were favored by natural (external) selection. Greater flexibility might be attained, for instance, by elongating the body, which might in turn depend on the kind of modularity available in the organization of the body axis, for example on evolutionary changes in the number of similar modules, as seen in annelids. Because new (duplicated) modules cannot occupy the same position as the original module, dissociation from the ancestral order of module appearance is likely and can in turn change the relationships of size or interaction. Thus, new serial modules can diverge in organization and then be co-opted for new functions (anisomerism). That is, it is feasible that all of these major changes could initially be driven by a behavioral shift anyway, as are most of the smaller morphological changes that can also lead to major macroevolutionary changes through a more gradual accumulation of transitions, according to ONCE (Fig. 1.2).

As argued by Sinervo and Svensson [339: 172], "Goldschmidt's hopeful monster… is in principle possible if mutations occur in regulative networks that buffer ontogenetic systems and yield functional integration; negative and positive regulation is common in endocrine systems and the organ systems targeted by hormones; one of the most famous natural examples of a hopeful monster, the axolotl, involves a loss-of-function mutation in an endocrine gene of major effect." Thus, "the likelihood of evolving hopeful monsters and morphological novelty is simplified if change is targeted to functionally integrated modules." As also noted recently by Dittrich-Reed and Fitzpatrick [114: 311], growing "empirical and theoretical research on hybrid speciation have revived the hopeful monster in a new, more credible form." Namely, the "recombination of parental chromosomes in the F2 and later generations during hybridization can generate genotypes that express phenotypes outside the normal range of variation observed in either parental gene pool, a phenomenon termed transgressive segregation." Transgressive hybrids often have higher fitness in new habitats, thus augmenting the likelihood of divergence from parental populations. They list a few examples of novel phenotypes that seemingly arose from hybridization: "extreme size of tiger x lion F1 hybrids; unique shapes and colors of hybrid orchids; ability of recombinant sunflowers to thrive in extreme habitats; specialization on a novel host plant in lonicera flies; and expression of novel gene transcripts (including new exons) via alternative splicing in hybrid poplars." As they recognized, "not all specific examples are relevant in nature, and not all would qualify as evolutionary novelty under certain definitions", but such cases serve to illustrate the "sudden appearance of profound differences between parents and hybrid offspring reminiscent of Goldschmidt's hopeful monsters." Also at a more microevolutionary level, in his book *Macroevolution—Patterns and Process*, Stanley [347] provided numerous empirical examples to defend the idea of "quantum speciation", referring to cases in which a very small population (for instance, of only 10 individuals) forms a new species, often very rapidly and often leading to "hopeful monsters".

Both Alberch's "Logic of Monsters" and Goldschmidt's "hopeful monsters" are deeply related to, and dependent on, the highly constrained and linked developmental mechanisms mentioned in Waddington's notion of homeorhesis and Gerhart and Kirschner's concept of facilitated variation. Moreover, rather than being opposing views of teratology, Shapiro's lack of homeostasis and Alberch's "Logic of Monsters" concern, above all, different degrees of abnormality as discussed in Chap. 7. Shapiro refers mainly to cases that are so severe that not only Waddington's normal homeostasis, but even homeorhesis, are perpetually disrupted with the resulting phenotype being more unorganized/unpredictable than in Alberch's "Logic of Monsters". Alberch refers to cases in which there is a disruption of normal development, but the strong constrained mechanisms involved in homeorhesis can still lead to a "logic" (order) even within abnormality. The same logic that can eventually lead to Goldschmidt's "hopeful monsters". In fact, Goldschmidt's notion of "phenocopies"—i.e. organisms in which experimentally produced abnormalities mimic certain mutants of the untreated organisms [147]—is very similar to Alberch's logic of monsters: induced defects are often seen in natural mutants as well as in variants of the normal population of other species or their ancestors.

Although I contend that there is often a "Logic of Monsters" in nature/evolution (see Chap. 7 and [106, 107]), I do not argue—nor did Alberch—that everything in life and evolution is "ordered". In fact, ONCE takes into account that many biological processes, features, forms, and behaviors are, for instance, significantly related to randomness (Fig. 1.2). As explained throughout this book, I am a strong defender of Gould's notions of contingency and randomness and of his "replaying life's tape" metaphor. In fact, one of my favorite examples used both to refute a gene-centric and an extreme externalist/adaptationist view of evolution—and to show the complex interplay between internal constraints, epigenetic factors, and randomness—was provided by a close colleague of Gould, Lewontin [228: 126]. As Lewontin pointed out, and as noted in Chap. 7, our faces are not totally symmetrical, nor are the facial muscles and nerves of each side of our head. However, the cells of the two sides of the face have the same genes and basically experienced the same external environment, so we need to include these examples of developmental epigenetic mechanisms, variation and randomness in evolutionary biology. That is, evolution is not explained by the genes, or the environment, or even the genes plus the environment.

Box—History: Platonic teleology, Aristotle, "Design", Intentionality in Nature, and "Bernard Machines"

In his book, *The Tinkerer's Accomplice—How Design Emerges from Life Itself*, Turner [360] developed an idea that has direct connections with ideas related to ONCE. Specifically, his idea links the occurrence of directional evolution leading to an apparent "design" of organisms with behavioral choices, niche construction, and homeostasis in a similar way to Gerhart and Kirschner's concept of facilitated variation. Turner provides a succinct historical background on the idea of intentionality in nature, referring, for instance, to platonic teleology, in which crabs were seen as made by God to offer particular moral examples to teach humans. Aristotle saw purpose in a different, more dynamic way, more like Turner does: that is, with a focus on physiology. For instance, according to Aristotle an animal is supposed to do something and, when taken out of that context, say a fish taken out of water, it will "intentionally" try to go back to its original context (water) or to adapt to the new context (e.g. land).

According to Turner, a major problem of Darwin's theory is "randomness". Turner proposed to solve this problem with his notion of a "directed goal" linked mainly to a physiological homeostasis that is somewhat similar to the idea of "facilitated variation". He used termite nests as an example of "Bernard Machines"—named for the French physiologist Claude Bernard, who first considered homeostasis as a central feature of living systems—and thus of a "design" that is not really planned. Instead, this apparent "design" is the outcome of a general homeostasis in which no single termite individual understands the functioning of the whole nest. Turner therefore tries to differentiate self-organized systems such as rivers versus "designed" biological systems. He refers, for instance, to the apparently more self-organized arterial system seen in early human fetuses versus the occurrence of a heart and numerous blood vessels to main organs as seen in human adults, which in his opinion requires "some sort of 'self-knowledge' by biological agents." He thus contrasts his idea with that of self-organization postulated within the physicalist framework of Newman: for him, Newman's concept of self-organization is in essence a Neo-Darwinian atomist explanation for "design".

However, although some of Aristotle's ideas were similar to Turner's—e.g. according to Turner the "soul", or "interconnection of parts", is the ultimate Bernard machine—Aristotle used ways to explain morphology that were unlike Turner's ideas [223]. One way is teleology, i.e. natural entities have intrinsic purposes, and the second was matter, which has some resemblance to the ideas of Newman's physicalist framework. Moreover, in a sense Aristotle also accepted some notions of vitalism [223]. Turner [360], in contrast, provided examples of how epigenetics can affect subsequent generations, not genetically but exclusively epigenetically. For example, bacteria are extremely important for our gut, which can experience changes in size and shape during ontogeny due to shifts in the microbial flora, which can in turn be passed on to the next generations by life birth, from mother to child. Turner also provides a fascinating, but very different, example showing how injured horns in mammals can have an abnormal growth, and when the horn is replaced in the following year, there is a "memory" in the sense that the new horn matches accurately the shape of the last, injured horn. However, Turner does not see this as an obscure vitalistic phenomenon. Instead, he uses it as an illustrative example of "self-knowledge". He refers to another fascinating phenomenon: that of the whiskers ("moustache") of mammals. These are represented by a "map" in the brain that is similar to their real phenotype, and when one whisker is removed from an animal, its nerve connections degenerate, and its corresponding portion on the brain may also ultimately vanish. Thus, the body has a "memory" of events that have happened during life (epigenetics), and this "memory" can affect the phenotype of the animals later in life.

However, Turner does not clarify, in these examples, whether such a "memory" could persist in subsequent generations. Do the progeny of the mammals with injured horns ever grow horns with the injured appearance? I predict that this is not the case, but if it were, it would obviously be a major vindication of Lamarck. The question of whether there is or not a body "memory" and, if that is the case, of whether it can or not be inherited, is related to Turner's idea that there are "persistors" in nature, which could have some kind of connection with the notion of behavioral persistence in Baldwin's organic selection. However, Turner described a "persistor" as an agent that can see/move toward the future, in contrast to the "replicators" of Darwin (e.g. genes), which can merely replicate the past. An example he provided concerns our blood, which is a new environment totally created by animals—by way of homeostasis—where rules of existence are very different from those of the external environment.

This notion of homeostasis as a process that profoundly separates organisms from non-living objects—in the sense that they can maintain their characteristics in the face of many, and sometimes very adverse, changes in the external environment—is one of the tenets of systems biology. However, Turner's repeated use of teleological concepts, such as "future" and "intention" sets his main ideas—or at least the way in which he expresses them—quite far apart from ONCE. This is because ONCE rejects teleological terms such as "future" except if they refer explicitly to organisms that, like humans—and apparently other taxa such as other primates, dogs and elephants [75]—can indeed consciously plan for the future and make behavioral choices according to their plans. But even if I think that we will recognize that such potentialities will be shown to be much more common in nature than it is normally accepted, as stressed by De Wall [75], they will still be probably the exception in nature: as explained in Chaps. 1, 2 and 3, all organisms can make behavioral choices even if most of them do not have the mental capacities and consciousness to plan for the future. And, although I do admire his 2007 book—and accordingly I am discussing it here in detail-Turner also does not address the issue of long-term directional macroevolutionary anatomical trends. I also disagree with Turner's use of the word "intention" in Bernard's Machines and termite nests. In my view, there is normally no "intention" at all, of any animal or agent, to "build" the whole machine as Turner recognized himself. However, apart from the use of such terms—and from the fact that he does not address the issue of long-term directional macroevolutionary anatomical trends—some of his ideas are clearly somewhat similar to those in ONCE, and I do think that his book helps to bring organismal biology back to evolution by stressing in particular the importance of physiology, which is often neglected in evolutionary biology.

In his book, *The Shape of Life*, Raff [302: 367–370] explains how we can emphasize the importance of epigenetics and niche construction in cases such as those of the termite nests without needing to invoke any notion of "purpose" or "intention" by the termites themselves (a point in which I completely agree: see previous box). Rather, the characteristic homeostasis of the termite colony can be explained by a complex combination of non-purposeful behavioral choices and epigenetic signals. For instance, experiments indicate that a rich set of social interactions exist between termites and that in some termite taxa contact between workers and the anal end of reproductives is essential for the transmission of some chemical substances that control differentiation. Both male and female reproductives manufacture a pheromone that represses reproductive differentiation, and soldiers likewise generate a pheromone that prevents workers from differentiating into soldiers; once soldiers are removed, workers start to become pre-soldiers.

A similar example provided by Raff concerns naked mole rats, which are diploid and develop as males and females, and actually can be also considered eusocial animals because they form a huge colony with tunnel systems that may expand up to 3 km and include more than 250 inhabitants. The colony has a single reproductive queen at any time, who prevents other females from entering estrus through a combination of bullying and pheromones. When she dies, other females become reproductively competent and fight to the death until a new queen is established. That is, as in termite colonies, the roles of individual animals in the naked mole rat colony change because of the behavior/death of and/or signals from others by way of epigenetic factors. Therefore, Raff [302: 370] emphasizes that the reproductive female in such social species goes "beyond the tissue or organ grades of organization to produce extracorporeal selves; the reproductive individual controls (epigenetically) the biological state and behavior of these developmental extensions of itself by pheromones, analogous to the hormonal regulation within the body."

A final example of a how concepts similar to a "facilitated variation", and recent knowledge of genomic networks, allow us to remove the need for teleological explanations was presented in Andreas Wagner's [385] book *Arrival of the Fittest: Solving Evolution's Greatest Puzzle*. The example concerns the transcriptional regulator that blocks the use of lactose, which binds to lactose when lactose is present, but that changes its form in the absence of lactose, thus becoming incapacitated to bind to the DNA to block the use of lactose. This is not a random event or mutation: it is instead directly related to the presence of a substance in the environment/niche. The behavioral shift by humans to domesticate milking animals and then to choosing to drink the milk—a behavior that persisted for generations through phenomena such as social heredity, and that persists today—was therefore a key factor in the whole process (for passionate discussions on details about which factor was more/less important regarding this specific example, see Weber and Depew's 2003 book).

In summary, on the one hand ONCE can explain—by way of behavioral persistence combined, for instance, with the subsequent accumulation of random mutations favored by natural selection within the niche constructed by that behavior—the gradual evolutionary changes seen in microevolution, as exemplified by this lactose example, as well as in major, long-term macroevolutionary trends as the ones illustrated by the titanothere horns (Fig. 4.1). On the other hand, ONCE can also account for saltational evolution, via for instance, Gerhart and Kirschner's facilitated variation, Waddington's homeorhesis and/or Goldschmidt's "regulation", which can be for example related to the existence of very intricate, and at the same time flexible, genetic networks [385]. I will further discuss these topics, and integrate them with the other subjects addressed throughout this book, in the chapters below.

Chapter 9
ONCE Ideas Are Put Together: Evolutionary Behavioral Ecology, Adaptationism, Systems Biology, and Interdisciplinary

In the previous chapters, I covered a wide range of different topics from numerous areas of biology and other scientific fields—from Aristotle's discussions about whether or not plants have a "soul" to mathematical network theory employed in anatomical network analyses—that could be seen as disconnected from each other and from the main subject of this book and from ONCE (Fig. 1.2). For instance, at first glance it might seem that the occurrence of both macroevolutionary trends and eco-morphological mismatches in nature is paradoxical and that these two phenomena are unlikely to be deeply related to each other. It might seem even more unlikely that the relationship between these phenomena can help to bridge the gap between seemingly contrasting ideas long defended by internalists and externalists, by microevolutionists and macroevolutionists, by Neo-Darwinists, Neo-Lamarckians, and Baldwin followers, and by current Evo-Devo lines of thinking such as those focusing on the physicalist framework, niche construction, and developmental epigenetic plasticity. It is therefore time to put everything together. I will start by summarizing how these topics are related and synthesize the overall, multidisciplinary vision of ONCE.

ONCE is an attempt to provide a sound explanation of such diverse and seemingly disconnected ideas and the links between them. This attempt is based on an extensive compilation of empirical data obtained by numerous researchers, both modern and ancient, from several diverse areas as well as on the reading and discussion of the broader theoretical microevolutionary and macroevolutionary ideas of authors such as Aristotle, Cuvier, Étienne and Isidore Geoffroy Saint-Hilaire, Darwin, Wallace, Lamarck, Baldwin, Osborn, Bateson, Waddington, Goldschmidt, Schmalhausen, Gould, Alberch, West-Eberhard, Newman, Müller, Galis, Wagner, Minelli, Pigliucci, and Odling-Smee, among many others. Importantly, the multidisciplinary view of evolution of ONCE provides a specific temporal dimension for the diverse phenomena highlighted by all these authors, including those that espouse the emerging Extended Evolutionary Synthesis (Fig. 1.3). This is because ONCE also points out how the timing in which each phenomenon—related to, e.g., internal constraints and internal selection, epigenetic factors, behavioral choices/shifts/

© Springer International Publishing AG 2017
R. Diogo, *Evolution Driven by Organismal Behavior*,
DOI 10.1007/978-3-319-47581-3_9

persistence and niche construction, random mutations, natural selection, and genetic drift—is more likely to be of particular importance within the transitions leading to events such as macroevolutionary trends and eco-morphological mismatches. ONCE thus helps to explain the apparent paradox between cases (1) of seemingly directional evolution—which have puzzled biologists since Aristotle and often led them to defend ideas such as those of scala naturae, progress, teleological purpose and design, or vitalism in evolution—and (2) of constrained evolution and behavioral shifts that, together with many other phenomena and also with chance, led to the much less appreciated and discussed occurrence of ecomorphological mismatches. This apparent paradox is related to the profound interplay between internal factors and factors influenced or selected by the external environment. In particular, ONCE stresses the fact that organisms are being selected, but also do the selecting, and that both types of selection are constrained by and related to a wide range of other factors.

During development, internal constraints related for instance to homeostasis and canalization, tend to constrain the diversity of/morphospace used by adult phenotypes but can also allow the existence/increase of developmental plasticity/hidden variation, while internal selection tends to constrain/decrease both the used morphospace and plasticity/variation (Fig. 1.2). In contrast, developmental epigenetic phenomena directly influenced by the external environment (e.g. related to hormonal regulation) tend to explore the available physiological, behavioral, and anatomical plasticity within the context of the specific niches/behaviors/ways of life that the members of the population occupy/display. Therefore, these phenomena have the potential to help direct evolutionary change, as also has genetic drift, for instance.

Behavioral choices/shifts of organisms are possible due to the plasticity resulting from and interplay between both internal factors and externally driven epigenetic events. When the first humans became bipeds, they had enough behavioral, physiological, and morphological plasticity to do so. Behavioral shifts are the main drivers of evolutionary changes such as those seen in macroevolutionary trends, in which the new behaviors are viable and followed by behavioral persistence through social heredity by way of phenomena such as teaching, learning, and imitation. For example, humans teach their babies to walk bipedally and babies imitate their parents when they try to stand bipedally: both the parents and the babies are active players in these processes. However, behavioral persistence is also linked to less emphasized phenomena such as those in which parents may directly or indirectly contribute to the abandonment, or even death, of their descendants if they, for instance, fail to learn/imitate/perform the new behaviors. Such behaviors are not usually found in modern humans, but are very frequent in non-human taxa, as explained in the previous chapters. Therefore, organisms are active evolutionary players that directly contribute to driving their own evolution and build their own niches.

Natural selection then comes into play as a secondary but likewise crucial evolutionary player. In the case of macroevolutionary trends, due to the behavioral persistence of the population, random mutations and/or epigenetic events leading to physiological, behavioral, anatomical, and/or genetic features that turn out to be advantageous within the context of the new behavior and niche and the external environment would be selected. For instance, during the development of organisms, epigenetic factors that sense and/or are directly affected by the external environment and/or the new niche/behavior can contribute to ontogenetic canalization in the same direction of the selected evolutionary trends. This effect further directs evolution and increases the match between behavior, phenotype, and external environment. Such phenomena further emphasize the importance of behavioral choice in the entire evolutionary process and its direct, strong links with cases of directional evolution because the behavioral choices of the parents and of other organisms within the niche lead to a continuous, interactive process that is interconnected in so many ways with the behavioral choices of the parents' descendants. That is why this process can extend for long periods of time, thus resulting in the observed macroevolutionary trends and, at least in the earlier stages of these trends, in a further increase in etho-eco-morphological matching leading, for instance, to cases of viable phenotypic overspecialization. However, even in these cases the existence of strong developmental internal factors causing phylogenetic inertia never allows an organism to be fully adapted to its ecological habitat/external environment, i.e. to reach an optimal etho-eco-morphological correlation.

Moreover, in many cases, behavioral persistence, loss of plasticity due to natural selection or to sexual selection, genetic drift, internal constraints, and/or simple chance, separately or combined with each other and with many other factors typical of, e.g., overspecialization, can make it difficult for the organisms to respond to new changes of/challenges by the external environment. This in turn may lead to etho-ecological, eco-morphological, and/or etho-morphological mismatches and potentially to cases of extinction. In addition, because of internal developmental constraints resulting from the fact that organisms are composed of many developmentally closely interconnected parts, in at least some cases natural selection of a certain trait will often result in correlated changes in other traits, which may be detrimental with respect to the direction of that selection. Emblematic cases of both etho- and eco-morphological mismatches due to internal constraints are the presence of hindlimb elements in whales and the abnormal occurrence of tails in human adults.

On the other hand—in what may seem to be a further paradox but rather shows instead the profound and complex interconnection of all these phenomena—internal constraints can themselves be crucial for maintaining some of the original phenotypic (e.g. morphological or behavioral) plasticity. For instance, plasticity is present at least in earlier developmental stages as hidden variation, which might allow organisms to display new behavioral shifts or revert to the ancestral, less specialized, behaviors and thus to escape eco-etho-morphological mismatches and/or evolutionary dead ends. Moreover, as a further example of niche construction and the central, active role played by organisms in evolution, the likelihood of a taxon reaching an evolutionary dead end is deeply related to the behavioral choices that helped to construct the niche in which the members of the taxon now live. For instance, it has been empirically shown that taxa living persistently in less complex/more stable environments such as caves tend to be taxonomically and behaviorally less diverse than those that colonized more complex/rich/unstable environments [4].

In a metaphor that summarizes the tenets of ONCE, organisms and their behavioral choices are the main drivers by selecting which road to take. Behavioral persistence—

associated with Waddington's homeorhesis, Gerhart and Kirschner's facilitated variation, and/or Goldschmidt's "regulation", externally driven epigenetic factors, and/or random mutations selected by natural selection, and various other factors such as genetic drift—is thus crucial to help direct evolution within the specific road that was selected by the organisms. This persistence thus plays a crucial role in delimiting that road as do the walls, trees, and mountains surrounding a highway. All of these factors, depending on the external environment, can thus contribute to the prolongation of this path (potentially leading to evolutionary trends) or to its blockage due to/leading to stressful conditions. The organisms might be able to respond to these stressful conditions with new behavioral shifts or instead a reversion to ancestral behaviors possible in great part due to internal factors that helped maintain the necessary plasticity to do so. Or they might not be capable to change their behavior at all, thus leading to evolutionary dead ends and eventually to extinctions.

This dynamic view of evolution, in which organisms are crucial active players due to their behavioral choices and persistence and niche construction, which will in turn affect natural selection and so on, is somewhat similar to the "two-way eco-evolutionary feedback loop" between organisms and their environments *sensu* Sultan [350: 158–160]. Specifically, she wrote: "when an organism modifies its environment in some way, it alters the selection pressures it experiences; this altered selective milieu can lead to evolutionary change in the organism, which will in turn change its subsequent impact on its environment, this change then leading to further evolutionary response." An example she uses to make her point concerns human evolution. In agriculture's early history, West African individuals who cleared lands to farm crops, such as yams, accidentally gave rise to favorable breeding for mosquitos, which then led to higher rates of malaria infection. The impact of this human behavioral choice, as well as subsequent behavioral persistence across various generations, in turn led to selective shifts within human populations inhabiting these malaria-endemic regions, such as an increase of the frequency "of the HbS or 'sickle cell' allele, as well as other blood cell variants that contribute to malaria resistence."

Importantly, as explained previously, putting organisms themselves, in particular their behavioral choices and shifts and persistence—which are linked to a trial-and-error process, to more complex cognitive behaviors in some taxa, and also simply to chance events—at the very center of biological evolution leads to an almost endless number of possible outcomes. For instance, an organism-centered perspective explains why etho-ecological, etho-morphological, and eco-morphological mismatches are so common, and probably even the rule, on this planet. The last example of mismatch I will provide here concerns what, in a way, can be considered to be the more extreme form of mismatches: those taxa or populations that create conditions that actually promote their replacement by members of other clades. One of the several such cases reviewed by Sultan [350: 104] is that the "deep shade cast by pine trees of the north temperate region" provides "too little light for the growth of pine seedlings but just the right conditions for the shade-tolerant seedlings of several broad-leaf deciduous trees." As noted by her, "it is this type of self-suppressing habitat modification that provides the motive force for ecological succession."

The issues discussed here thus have implications not only for many areas of biology and science in general but also for the understanding of our own evolution and path specifically. This is because, in a sense, humans seem to be dangerously stuck in a persistent, monotonous behavior in which they do not seem to be flexible/plastic enough to voluntarily give up/decrease/successfully control their investments in technology as noted previously. Unfortunately, this is leading us to a scenario in which technology, consumption patterns, overpopulation, and pollution are beginning to seriously compromise the integrity of the global ecosystem and thus of our own existence.

A main problem with evolutionary thinking is that many authors focus on what Matsuda [245] called the "ultimate" evolutionary processes, i.e. those processes in which natural selection helps direct evolutionary changes after the behavioral choices/shifts—the "proximate" evolutionary processes *sensu* ONCE—occur. For instance, a certain population of pearlfishes live parasitically inside sea cucumbers, and natural selection is thus crucial to select those random mutations/epigenetic phenomena that provide advantages within that parasitic mode of life. However, this is putting the cart before the horse because the first, crucial event that initiated the whole process was the choice of those pearlfishes that originally entered and stayed inside the sea cucumbers. As pointed out by Matsuda, parasitism almost always involves organic selection because the choice of the hosts is made by the organisms that will become parasites just as sexual selection is organic selection because the choice of the mates is made by the organisms themselves (Chaps. 4 and 5). In this sense, the key tenet of ONCE—i.e. that natural selection is still a crucial player in evolution but is mainly a secondary player, particularly in terms of order of events (Fig. 1.2)—is similar to that of authors such as Odling-Smee et al. [276: 179], who stated that niche construction, including behavioral choices and shifts, is pro-active "while natural selection is reactive".

As its name indicates, natural selection mainly "selects" among already existing features and thus forcibly decreases diversity and complexity, whereas Baldwin's "organic selection" often refers to cases in which there is a direct increase of diversity, e.g. when an individual from a taxon makes a behavioral choice that was never made by any other organism. That is why I prefer to use the term "organic evolution" instead of "organic selection" as explained in Chap. 1. Phenomena related to randomness also tend to increase complexity and diversity as empirically shown, for instance, in phenotypic studies of *Drosophila* mutants [256]. However, this does not necessarily mean that natural selection can only have a "negative" constraining role in evolution *sensu* Gould. As explained previously, according to ONCE natural selection is crucial, as a secondary player and linked to phenomena such as behavioral persistence, in directing evolution in cases of e.g. long-term macroevolutionary trends, including those that lead to major morphological novelties (Fig. 1.2). Pure randomness by itself would almost never lead to cases such as the huge horns resulting from the evolutionary trend seen in Fig. 4.1 or to the numerous different features related to bipedalism that our ancestors acquired in just a few millions years. Saying that natural selection is somewhat similar to the walls and/or trees around a road could imply such "negative" constraints. Instead, this metaphor is intended to show that without those walls and trees one could never travel, for example, to far away roads that lead to highly modified forms/novelties. That is, this metaphor stresses not only the "negative" but also the "positive" driving force of natural selection.

In Chap. 8, I explained that many Evo-Devoists and followers of the Extended Evolutionary Synthesis follow the motto "evolution as the control of development by ecology". This motto has the strength of taking into account ecology as a major component of evolution, but in my opinion it still does not focus enough on the behavioral choices made by the organisms themselves or on the active, key role of the organisms in their own evolution. In this sense, this motto is also far from some of the current ideas of the emerging field of Evolutionary Behavioral Ecology. This field, and particularly the 2010 book with the same title edited by Westneat and Fox, is highly relevant for broader discussions about ONCE, and I will discuss it here in some detail here. Unfortunately, just as the Extended Evolutionary Synthesis largely ignores behavioral ecology, that book also ignores some of the points emphasized by this synthesis and even by current Evo-Devoists in general. That is, although behavioral ecologists do ascribe a more active evolutionary role to organisms, they continue to be highly influenced by both the "struggling" view of evolution and, in at least some cases, an extreme adaptationist Neo-Darwinian framework that moreover is often associated with the use of teleological terms (see box below).

Box—Details: Evolutionary Behavioral Ecology, adaptationism, and teleology

The influence of adaptationism and teleological thinking in Evolutionary Behavioral Ecology is well illustrated in the first chapter of the book with the same name: "behavioral ecology is largely theory driven… about what animals should do in particular circumstances to maximize the fitness benefits" [38]. Another example is the following sentence, which is also in that chapter: "the sequencing of the honeybee genome revealed a low number of immune genes relative to nonsocial insect species, suggesting that bees have developed social behaviors to deal with disease and pests." It is one thing to use—as behavioral ecologists do and as done in ONCE—the term behavioral choice in a way that does not imply necessarily conscious decisions; this is not teleological at all. However, it is a completely different thing to state that organisms can plan ahead how "to deal" with "disease" and "pests" when these organisms clearly do not know what "disease" and "pests" are. This is teleological thinking, and in my opinion it is the wrong way to explain what is happening. Honeybees did not plan ahead to deal with disease and pests. They simply made behavioral choices related to their day-to-day tasks, which in the long term, either by chance (e.g. as a byproduct) or more likely due to natural selection within the context of the niches constructed by those behavior choices, ultimately led to better protection against pests and diseases.

Within the same book, Fox and Westneat's [133: 29] chapter on adaptation is a particularly strong, and I would say extreme, defense of the adaptationist framework. They explain that "Gould and Lewontin objected to the adaptationist paradigm adhered to by most behavioral ecologists", but defend that this paradigm "remains dominant in behavioral ecology because, in case after case, the focus on adaptationist explanations has led to new insights." However, "leading to new insights" does not mean that one is following the right scientific path. Everything, even non-scientific reasoning, tends to lead to "new insights". Although I do not in any way intend to place Intelligent Design and the adaptationist paradigm on the same level, one can also say that the biased discussions of followers of Intelligent Design also have led to many "new insights". This fact is evidenced by the endless number of new books and articles and blog posts they continue to write and even museums they build. They continue to write so much, and always confirm their *a priori* ideas, precisely because they blindly follow a path and a way of thinking that is often difficult to falsify, and even when it can be falsified they ignore the data that do so. This can lead to a dangerous circular reasoning of "predictions", "new insights" and "confirmation" of (the often untestable) predictions, and so on.

This circular reasoning can be shown with an example used by Fox and Westneat to support their defense of the adaptationist framework. The authors state that "a perfect example" of such new insights was "David Lack's hypothesis that clutch size in birds would be optimized to balance the number of offspring produced with the parent's ability to feed those offspring well enough to survive." In other words, "parents should produce a clutch of a size that maximized the number of offspring fledged from that clutch." As they recognize, "experimental studies on multiple species of birds revealed that clutch sizes were close to, but did not match exactly, what Lack predicted; most birds produced slightly smaller clutches than predicted." But, they explain, because "Lack was invested in the adaptationist paradigm", "despite the possibility that many nonadaptive hypotheses could be proposed to explain the disparity between data and theory, Lack chose instead to hypothesize that other factors affected selection on clutch size." According to their own words, "this search for adaptive explanations led to a diversity of new adaptive hypotheses." In particular, "many studies show that parental work load is indeed important in lifetime reproductive success." Therefore, they concluded that "no doubt some nonadaptive processes also affect clutch size in birds, but Lack's focus on adaptive processes nonetheless led to substantial new insights." That is, one has a paradigm (adaptationism) and then makes predictions exclusively within that paradigm, and then even when there is a "disparity between data and theory", one does not question that paradigm but instead continues incessantly to make new predictions, until the paradigm is "proven", without even questioning whether the problem might be the paradigm in the first place. Of course, the excellent works of Fox and Westneat show that they do not really employ such circular reasoning in their studies; I am mainly debating here their use of words and examples in that particular chapter of their 2010 book.

Within such an adaptationist framework, it is not surprising that in the whole *Evolutionary Behavioral Ecology* book the occurrence of etho-ecological, etho-morphological, and eco-morphological mismatches is almost completely neglected, because it does not fit into that framework. Even in those cases in which the authors recognize that empirical data show the existence of such mismatches, they often refer to them as exceptions related to particular cases of "personalities" or "syndromes". In their chapter, Ghalambor et al. [146: 94] state that, aside from conditions that select for nonflexible or flexible behavioral choices, plasticity may also be constrained by various factors, e.g. because "behavioral traits are inter-correlated within individuals, a phenomenon manifesting as personalities or syndromes." That is, "personality describes the limitation of behavioral plasticity... thus, although it might benefit animals to be less or more aggressive in specific contexts, selective history and development may constrain individuals in how plastic they can be, via their personality." The authors do recognize, however, that "the existence of such syndromes has important implications for how selection acts on suites of behaviors and for the maintenance of behavioral variation in populations." But in Nonacs and Blumstein's [274] chapter, the authors do emphasize that "behavior syndromes" can lead to maladaptive behaviors and thus to evolutionary mismatches. As explained in Chap. 5 of the present book, Lindholm [232] recently reviewed various empirical examples of etho-ecological mismatches in a wide range of taxa, which were accordingly designated "maladaptive behavioral syndromes" in that work.

Authors such as Holekamp et al. [189] also criticized disciplines related to ecology for being mainly based on the adaptationist framework. According to the authors, there are endless studies emphasizing how developmental plasticity and hidden variation can allow organisms to display new behavioral shifts to respond "adequately" to environmental changes. However, many fewer studies have addressed the opposite scenario: how developmental constraints can severely limit the behavioral flexibility of organisms and their responses to changes in the environment, thus potentially leading to etho-ecological mismatches. Holekamp et al. argue that developmental constraints do affect variation in behavioral flexibility among individuals and also among species and higher clades, and they review several empirical examples to support their view. For instance, prenatal hormone exposure constrains the flexibility of aggressive behavior in individual spotted hyenas during their entire lives by influencing their baseline aggressiveness. They note that such hormonal epigenetic effects are often not taken into account in socioecological models of mammalian behavior, which suggest that aggressiveness should instead be shaped adaptively by resource competition. Among spotted hyenas, resource competition is almost always more intense for low- than for high-ranking individuals, but—contrary to what would be expected according to such adaptationist models—the relationship observed between prenatal androgen exposure and aggressiveness occurred independently of social rank.

Another example concerns a mix between functional and developmental constraints. The fact that mammals such as carnivores have paws instead of hands like ours drastically limits their tactile interactions with their environments, thus in turn limiting their social interactions when one compares them with, e.g., those of primates.

Moreover, Baldwin and his idea of organic selection, which stresses the crucial importance of behavior, are almost completely neglected in the same 2010 book. This omission seems strange in a book dealing with the evolution of behavior, but is very likely related to the extreme Neo-Darwinian view of evolution referred to previously. In the entire index of that book, which contains hundreds of entries, there is a single entry for Baldwin, and furthermore it concerns the Baldwin effect rather than his broader ideas about organic selection. Moreover, as odd as this may seem in a book entitled *Evolutionary Behavior Ecology*, a clear example of gene-centered ideas is defended: "behavioral differences that persist... suggest a genetically heritable basis" [335: 73]. It is difficult to understand why behavioral ecologists would assume that behavioral persistence should necessarily be associated with genetic fixation. There is no evidence at all to suggest that, for instance, the behavioral persistence (culture) of certain groups of primates to use stones to crack nuts, or of Portuguese people (like me) to eat salted cod so frequently, is necessarily "genetically fixed". In fact, as noted in Chaps. 1–3, the so-called Baldwin effect—i.e. the hypothetical genetic fixation of behavior traits—was, within the bigger scheme of Baldwin's organic selection, the exception rather than the rule in behavioral persistence, which was mainly related to social heredity—or, using more modern Evo-Devo concepts, to ecological/behavioral inheritance (Fig. 1.3).

My intention is not to criticize the discipline of Evolutionary Behavioral Ecology per se or, for that matter, disciplines such as socioecology. In particular, I surely do not intend to criticize the book *Evolutionary Behavioral Ecology* [335]. That book is hugely important as a source of examples of remarkably interesting behavioral traits. Also, as stressed previously, even in the aspect of the book that is weaker in my opinion—the evolutionary Neo-Darwinist ideas and adaptationist paradigm followed by various of its authors—my criticism only applies to some chapters/parts of an otherwise excellent book. For instance, Ord and Martins (2010: 126) do emphasize that some cases of behavioral persistence can lead to situations in which a certain behavior no longer serves "current adaptive function", i.e. that it can lead to etho-ecomorphological mismatches, a key assumption of ONCE. The authors also refer to the case of cowbirds and cuckoos, birds that lay eggs in the nests of other bird species to avoid the considerable investment needed to raise their progeny, whereas some of their target species evolved ways to recognize and throw out the foreign eggs from their nests. However, surprisingly, Californian loggerhead shrikes display a similar behavior when fake eggs are experimentally placed into their nests despite the fact that these birds are not currently parasitized by any other birds, meaning "that this rejection behavior has no (current) function". They note that phylogenetic studies show that loggerheads manifest this behavior because they inherited it from ancestors that seemingly were parasites and not because it serves any adaptive function today. That is, this would be a case of phylogenetic inertia affecting the current behavioral choices of an existing species, very likely by way of continuous behavioral persistence, exactly as postulated by ONCE.

Moreover, the book *Evolutionary Behavioral Ecology* provides a rich array of examples that clearly contrast with the one just mentioned, thus showing how diverse and complex—and how versatile and yet constrained—behavioral evolution can be as assumed by ONCE. In particular, the book provides several examples of behavioral shifts that reveal how organisms can be particularly behaviorally alert and how this can often lead to new behaviors/behavioral shifts. For instance, in their chapter, Ghalambor et al. [146: 105] reviewed experimental studies that increased versus decreased the risk of nest predation when birds selected sites for nests. The birds reacted plastically by either choosing more hidden sites or changing their nest sites to areas they perceive as safer. In turn, such behavioral choices can lead to a cascade of other behavioral changes, thus emphasizing how evolution can be strongly directional, and occur relatively rapidly, when it is driven by behavioral shifts. For example, some of the birds that shifted to more concealed nest sites in response to elevated predation risk subsequently visited the nest to feed their young more often than those that did not display the shift. As stressed by Ghalambor et al., "such plasticity suggests the cognitive ability to both monitor ongoing risks and opportunities and calculate the relative merits of each behavioral decision."

Importantly, that book also stresses that such cognitive abilities, as well as cognitive learning, are not limited to vertebrate taxa, as has also been argued throughout the present book. For instance, there is evidence from numerous species that previous experience affects behavioral choices, e.g. in fruit flies mating with a male of any size occurred more frequently after the female was courted by a small rather than by a large male [177: 169). In addition, males that had experience with recently mated, disinterested females sped up the beginning of courtship with virgin females, and those that had experience with immature disinterested females approached adult females more rapidly. Healy and Rowe [177: 171] also provide empirical evidence of the links between behavioral choices, learning, and epigenetic factors/changes, which are key for ONCE (Fig. 1.2). As an example, experimental studies in birds pointed out that experience with food storing causes increased neurogenesis and leads to enlargement of the hippocampus. Birds that store food recall what they have stored, when and where they did it, and also remember food locations over longer times than non-storing species can. According to the authors, hypothetically greater selection pressure on spatial cognition in food storers also lead to a specific augmentation in their spatial cognition: damage to their hippocampus results in difficulty with accurate retrieval but does not reduce motivation nor capacity to retrieve food locations specified by learned color cues.

Moreover, as noted by Sultan [350: 120–136], ecologists and in particular community ecologists are now recognizing increasingly more that "per capita trophic, competitive and other ecological effects are not constant" but "vary as a result of the behavioral, morphological and life-history adjustments made by interacting individuals in response to each other and to other aspects of their environments." This reinforces once again the fact that organisms are not mere automata in an ecological fixed context in which one can predict using mathematical equations what will be the balances between the number of predators and prey and so on. In fact, this is my single major concern with Carroll's [54] recent book, *The Serengueti Rules*. That book clearly has noble intentions, particularly calling attention to how humans are destroying this precious, marvelous planet. However, it does propagate a mechanistic view of organisms by explicitly arguing that, just as in our body we need specific ratios of e.g. substance/protein A versus substance/protein B, there is also a fixed limit in an ecological context, such as that of the Serengueti region in Africa, between the ratio of the number of certain predators versus the number of certain preys. However, as explained by Sultan, the effect of a predator population on its prey will be different if, for example, individuals alter their behavior so as to minimize predation, e.g. when fish-eating bass are present in ponds, potential bluegill behaviorally choose to move from more open water to littoral vegetation. That is, organisms are not simply like a certain substance or protein. Due to the phenomenon known as *emergence* in systems biology (see Chap. 1), they undertake complex behavioral choices, which are often unpredictable, and the Serengeti provides many examples of that: after millions of years of human evolution, animals such as lions, hippopotamuses, and snakes continue to be unpredictable enough to be able to surprise and kill members of our species.

A clear illustration of how ecologists had historically seen organisms as mainly predictable automata—seemingly without any type of ability and/or drive to undertake behavioral choices—within a community ecological context concerns studies on plants and pollinators. Specifically, as reviewed by Sultan, simulation models had "assumed that pollinator behavior would remain constant in the event of changes to a community's set of floral visitors." Once again, empirical data contradicted such a simplistic assumption: "a manipulative field study found that loss of even a single insect species led to major behavioral shifts by remaining pollinators." Namely, "in response to the removal of the most abundant pollinator species, the remaining bees immediately changed their behavior: they reduced their fidelity to particular plant species and dramatically increased their foraging visits to plants of other species, even within a single foraging flight." Therefore, as stressed in Sultan's book, one needs to have a broader interdisciplinary view of life and evolution that is less mechanistic and more dynamic because "the community impact of competitive, trophic, or other interactions generally reflects both (1) the density of interacting populations"—as traditionally pointed out by ecologists and as reflected in Carroll's 2016 book—"and (2) trait adjustments by individuals", including behavioral choices and persistence as argued in ONCE.

The reason for including the previous paragraphs, boxes, and discussions in this chapter is that they thus provide, in my opinion, a powerful illustration of one of the major strengths of ONCE: it integrates phenomena that are used by different authors, from internalist Evo-Devoists to behavioral ecologists, to defend different and supposedly opposed/contrasting views of evolution, e.g. gradualism versus saltationism, internalism versus externalism, gene-centered versus epigenetically-centered, structuralism versus functionalism, and so on. Other authors have recently stressed that such dichotomies, for instance externalism versus internalism, are often exaggerated and that reality is more often grey than black and white. For example, Lowe [238] stated that one of the key problems with such dichotomies is that they limit the kinds of variation that are critical to understand the causes implicated in the production of ontogenetic outcomes. If, for instance, the constitution of microbiotic communities in organisms (e.g. in human guts) "is conceived to have an environmental (external) source, the role of the microbiome in development and health demonstrates that the instructive-permissive distinction as an heuristic means of identifying which variation is methodologically and theoretically important, is deeply flawed" [238: 459].

As an example of such dichotomies and contrasting views, Moore [263: 119] presented a short, elegant summary of features that have been revealed by Evo-Devo—e.g. the importance of canalization, constraints, and externally driven epigenetic factors—and emphasized how these features are significantly different from the ideas of Neo-Darwinists. First, "if organisms are organized systems of multitude, developmentally interconnected parts, it will rarely be possible for selection to act on isolated traits". That is, "selection on a trait is very likely to bring about correlated changes in several others, which may be favorable, neutral or even detrimental with respect to selection." Second, "the functional demarcations of traits, which are so useful for discussing adaptation, lose usefulness in explaining development… because interactions frequently cross boundaries within developing organisms, and new functional capacities frequently develop from the raw materials left by functionally discontinuous earlier stages." Third, "both genes and environment become relevant to development only when they contribute in material ways to organismic processes, that is, at the cellular level, after they are translated or transduced into the molecules that cells use… questions about what genes do are likely to be raised in terms of developmental processes rather than programming or outcomes… the same is true for questions about the environment."

A further argument for ONCE is that the data available about the order of appearance of genetic, behavioral, and phenotypical traits associated with evolutionary trends and/or novelties support its predictions. In fact, a major contribution of Evo-Devo has been to show that many of the genetic traits and/or pathways associated with the development of structures such as the eyes of insects and vertebrates were present in the ancestors of these groups well before the appearance of those structures in evolution. This is often explained by pleiotropy, i.e. the same gene can be involved in various pathways and/or the formation of various structures (genetic/molecular plasticity). And, as explained in the previous chapters, it is becoming increasingly accepted by many researchers, due to a rising number of empirical behavioral and ecological studies and the decrease of the historical bias created by Neo-Darwinism, that behavioral shifts often occur without, and thus *before*, any phenotypic changes.

As explained throughout this book, a key difference between ONCE and other main evolutionary frameworks—including Darwinism, Neo-Darwinism, Lamarckism, Neo-Lamarckism, Baldwin's organic selection, and Evo-Devo's Extended Evolutionary Synthesis (Fig. 1.3)—is that ONCE takes into account the very common occurrence of etho-ecological, etho-morphological, and/or eco-morphological mismatches in nature. Of course, one could argue that Baldwin, as well as Darwin and the Neo-Darwinists, knew very well that "optimization is about constraints and trade-offs, not perfection", as put by Seger and Stubblefield [332: 94]. However, as pointed out in Chaps. 1 and 2, Baldwin did believe in progress toward an optimum. Moreover, it seems clear that the idea of eco-morphological "optimization" espoused by Baldwin and the Neo-Darwinists is very far from what is seen in the rapidly increasing number of empirical cases showing evolutionary mismatches and/or instances in which phylogeny is a far better predictor of anatomy than is ecology. This is illustrated by the fact that most of the ecomorphologists who undertook the studies reviewed in Chap. 6—who were often expecting to obtain the opposite results—did not hide their surprise, and often their discomfort, with the results they obtained.

Concordantly, Crispo [64: p. 2470] stated that "maladaptive plasticity was not considered in Baldwin's theory." As she noted, Baldwin proposed that plasticity itself could be adaptive and selected for, thus increasing plasticity in a population. Plasticity should therefore often increase under selection if the most plastic individuals are the most able to colonize a new habitat or persisting in a fluctuating one. Crispo noted that some authors argue that plasticity itself would not be selected upon, but is instead the subject to indirect selection by way of selection on the most extreme trait values representing the upper limits of the plastic response. Other authors argue that selection can occur directly on plasticity if plasticity increases the matching between the habitat and its optimal phenotype.

However, as Crispo noted, although these two views of selection on plasticity differ, their outcomes are identical in the sense that plasticity will increase due to selection whether selection acts on the level of plasticity or on the induced traits/trait values. She then reviews several empirical examples indicating that plasticity is higher after selection under new environmental conditions. For instance, a study found that the body shape of pumpkinseed sunfish (*Lepomis gibbosus*) was more plastic in the derived open-water ecomorph than in the ancestral inshore ecomorph despite the fact that the habitat of the latter is more heterogeneous. Another study revealed that the head and jaw length of tiger snakes (*Notechis scutatus*) were plastic in island populations that feed on large prey but not in mainland populations that feed on small prey. These are the type of examples and line of thinking that Baldwin had in mind, and so he focused less on cases of maladaptation and/or decrease in plasticity that would lead to eco-morphological mismatches and ultimately to extinctions.

As noted by Gould [159] and explained in Chap. 7, developmental internal constraints should not be seen merely as "negative" factors limiting the possibilities of evolution. Arthur [17] followed a similar line of thought, but used different terminology, in his book *Biased Embryos and Evolution*, in which he mainly divided developmental factors into negative ones (constraints) and positive ones (drive). Developmental constraints are particularly important within ONCE because, for instance in cases of overspecialization that could lead a taxon/population to an evolutionary dead end, the plasticity/variation that exists precisely due to these constraints may be crucial to revert or change the direction of evolution, particularly when associated with heterochronic phenomena (Chap. 7). ONCE thus provides a new way of linking behavioral shifts and persistence, epigenetics, random mutations, natural selection and internal factors—such as those related to Waddington's homeorhesis—that are associated with directional evolution and the loss of plasticity and potentially to overspecialization and eventually to extinction. In other words, internal constraints are crucial in maintaining some of the original phenotypic (e.g. morphological or behavioral) plasticity—at least in earlier stages of directional evolution—as hidden variation that might subsequently allow new behavioral shifts or revert to the ancestral, less specialized, behaviors and thus to escape evolutionary dead ends, as noted above.

A very interesting empirical example supporting this view, specifically the point that ancestral developmental constraints can facilitate new evolutionary possibilities, was provided by Rajakumar et al. [303: 80]. In a wild ant (*Pheidole morrisi*) colony, these authors discovered numerous anomalous soldier-like individuals that were notably larger than normal soldiers and, contrary to normal soldiers, displayed mesothoracic wing vestiges. These anomalous soldiers are similar to a super-soldier subcaste that is known to be continually produced in eight *Pheidole* species, further supporting Alberch's "logic of monsters" in which anomalies within a certain taxon are often mirrored in the wildtype of another taxon. According to Rajakumar et al., the similarity between the super-soldier-like anomalies in *P. morrisi* and the super-soldier subcaste suggests a developmental origin. Namely, the soldier subcaste is determined late in larval ontogeny at a soldier–minor worker switch point, which is mainly controlled by nutrition and mediated by juvenile hormone. The soldier-determined larvae grows larger than the minor worker larvae and forms a pair of vestigial forewing discs in their mesothoracic segment. These discs show a soldier-specific expression of spalt (sal), a key gene in the network underlying wing polyphenism in *Pheidole*, because its expression is spatiotemporally linked to the induction of apoptosis in these vestigial forewing discs.

Thus, according to the authors, the super-soldier-like anomalies found in *P. morrisi* likely originated from the abnormal growth of soldier larvae and their vestigial wing discs, i.e. by way of ontogentic changes that lead to these two features. That is, the potential to produce super-soldiers was probably present in the last common ancestor of all *Pheidole* species that evolved approximately 35–60 million years ago and has been retained since then in species of this genus, most likely due to internal constraints. Importantly, this example also shows how such constraints not only facilitated parallel evolution, per se, but also allowed shifts to ancestral or even new behaviors. In one of the eight species that continually produced super-soldier subcastes, *P. obtusospinosa*, the behavior of the members of that subcaste is different not only from that of other subcastes but also from the presumed ancestral behavior of any subcaste within the *Pheidole* last common ancestor. Specifically, the super-soldiers of that subcaste block the nest entrance with their extra-large heads and engage in combat to defend against army ant raids.

Rajakumar et al. also stressed that ontogenetic constraints can also be positive drivers of evolution because, as in this example and many others, they can be related to Waddington's notion of homeorhesis and Gerhart and Kirschner's notion of "facilitated variation". Because of constrained developmental interactions (e.g. genetic pleiotropy), mechanisms (e.g. associated with the patterning of skeletal–muscular connections in ontogeny) and genetic networks, a change in one single feature can result in a cascade of related/coordinated events that may allow for saltatory evolution and in turn facilitate even more homoplasic evolution. I do not fully agree with Reiss [305] that the term "constraint" is necessarily teleological because it gives the idea that there it is a defined goal that is being constrained. That was never the intention of Gould as exemplified by his idea that in many cases constraints can be crucial "positive" factors leading to major evolutionary changes. A clear empirical/phylogenetic example is the cascade of morphological characters resulting from a single major change (paedomorphosis) that repeatedly occurred in various salamander taxa [e.g. 401]. In such cases, it is likely that many phenotypic features that come with the "paedomorphosis" package are not advantageous (their persistence over generations being thus an example of evolutionary "trade-off" or by-product), and that some are even disadvantageous within the current behavior of the organisms (etho-morphological mismatch) and/or the habitat in which they live (eco-morphological mismatch). This is one more reason why one should avoid obsessively looking for optimizations/adaptations, in each and every morphological trait of each organism, for their current way of life.

As explained previously, epigenetic factors, combined with behavioral persistence related for instance to learning and associated with natural selection—and to "positive" internal constraints such as those mentioned in the previous paragraphs —can also contribute to directional evolution. This is because they can allow organisms to "sense", and be influenced by, the external environment during their development. In addition to the examples given in Chaps. 7 and 8, other empirical case studies have revealed a complex variety of proteins and chemicals that regulate the activation and transcription of genes, thus assembling what now is known as the epigenome. For instance, one epigenetic phenomenon, epigenetic regulation, controls cell differentiation, keeps the stable phenotype, and accommodates the organism to its local environment by way of histones and methylation of selected DNA sequences [232]. As noted by Lindholm, cases of transgenerational and persistent epigenetic inheritance propagate, thus pointing to selectable traits mediated by extra-genomic information. For example, phenotypic differences between humans and chimpanzees are mostly epigenetic rather than genetic; diversification of common structures is partially led and constrained by shifts in signaling pathways during early ontogeny [232].

Authors such as Bateson et al. [33] even prefer to use the term "Predictive Adaptive Response" to explain some epigenetic factors influenced by the external environment although, as stated elsewhere in this book, one should be careful with the extreme panselectionist adaptationist context in which some of these ideas are being developed in the field of Evolutionary Medicine. According to these authors, environmental induction can provide a "forecast" about the future environmental conditions that the organism will experience. In mammals, such a "forecast" would be mainly done via the mother. For example, vole pups born in the autumn have much thicker coats than those born in spring, the cue to produce a thicker coat being related to hormonal signals acquired prenatally from the mother, depending on day length. According to these authors, the benefits of a match between predicted and actual environments vary from case to case. For instance, East African acridoid grasshoppers, usually green, morph into a black form after a savanna fire blackens the habitat; the grasshoppers' plasticity enables them to form, even during adulthood, a camouflage that decreases the risk of being eaten by birds. Returning to the discussion about Lamarck (see Chap. 5), the authors provide several examples that further point out that not there is not only behavioral inheritance and ecological inheritance but also other types of epigenetic inheritance *sensu* them and the Extended Evolutionary Synthesis (Fig. 1.3) and *sensu* ONCE (Fig. 1.2).

Lindholm [232] reviewed empirical evidence supporting a profound connection between epigenetic factors and behavioral persistence. Pup licking and grooming in rodent mothers causes higher densities of glucocorticoid receptors in the central nervous system of the progeny. This phenomenon is linked to increased tolerance against stress, and the learned behavior has a heritable component in the sense that pups exposed to grooming subsequently tend to behave accordingly with their own descendants. Similarly, prenatal auditory song perceptions are necessary for later song learning and recognition of conspecifics in many birds. In mammals, initial dietary preferences are established prenatally when the fetus becomes familiar with the diet of the mother by way of the placenta. Pups also gain dietary information by way of scents from the mother's fur, mouth, and urine, and through lactation, as new flavors from her diet are mediated. Birds often feed on several different food items depending on habitat features and seasonal opportunities in which memory and learning are crucial. Importantly, Lindholm also reviews empirical data showing that food choice is affected by the explorative willingness of the organisms themselves. For instance, dietary-innovative taxa have higher speciation rates, likely due to their greater ability to explore new niches and habitats, as seen in blue tits that pierce through milk bottles left outside house doors to drink their cream, a habit that rapidly dispersed and was even adopted by other species, probably by way of imitation/learning. According to Lindholm, such learning-driven behavioral shifts open up new resources for the population but also expose them to changed selection pressures because mutations facilitating the behavior will be selected at the expense of individuals who need to learn from scratch, potentially leading to genetic assimilation *sensu* Waddington and/or to the Badwin effect.

An emblematic piece of empirical evidence of social heredity that was reviewed by Lindholm concerns preferences that individuals learn early in life, e.g. when anadromous fish return to their natal river or when ovipositing insects go back to the area where they hatched. As also stressed by him, increasingly more evidence is demonstrating that habitat choice of one species affects the choices made by other taxa. For example, researchers installed birdhouses for tits and then marked those that were occupied with white circles and the vacant ones with triangles. When migratory flycatchers later arrived, the researchers hung up additional birdhouses, now displaying both triangles and circles, but the flycatchers solely settled in birdhouses with circles, indicating that they had observed the brood differences of the two birdhouse types and chose the ones that appeared to be more successful. A more recent example reviewed by Lindholm is the ongoing urbanization of European blackbirds, a species ancestrally confined to forests. Some individuals invaded the city of Erlangen in Germany in 1820, and although this urban habitat differed greatly from forests, they were able to breed and interacted successfully with the foreign competitors, predators, and diseases. In fact, urban blackbirds rarely interbreed with the rural ones and are different from them genetically and behaviorally, in a further example of how behavioral shifts can easily lead to speciation and thus be potentially crucial for cladogenesis and evolution.

Similar views that behavioral changes are crucial for evolution are becoming increasingly more frequent in the literature as explained throughout this book. I think it is particularly worthy to refer here to the excellent book, *The Origin of Higher Taxa*, in which Kemp [209: 69] wrote, that "the first (evolutionary) stage is necessarily the behavioral one… followed by the developmental response of morphological modification." Kemp then stated that behavioral changes associated with phenotypic plasticity are probably related to many speciation events, but asked whether such events can also be significant in major macroevolutionary transitions leading to the origin of new phyla. He noted that authors such as West-Eberhard [393] suggested that they can do so, providing examples such as the origin of tetrapods, in which intermediate species with facultative air-breathing phenotypes would be capable of a plastic behavioral response to the oxygen concentration of the water. However, he then concludes that "apart from these rather special cases, there is no evidence that phenotypic plasticity plus genetic assimilation plays a role other than as a process of microevolution" [209: 70]. Still, Kemp then reviews examples, in other parts of his book, that do directly associate behavioral changes, major macroevolutionary events, and changes in body forms. For instance, in page 166, he states that—regarding the acquisition of cetacean (i.e. whale) characters—evidence exists that the "initial step, as represented by pakicetid-grade taxa, was extension of feeding (behavioral) habits into water by a still terrestrial artiodactyl." This is because at that stage "skeletal adaptations were limited to relatively minor dental changes and an ear capable of a degree of subaquatic feeding." This example, converning one of the most emblematic macroevolutionary transformations in body form across vertebrates, is therefore a further case of major behavioral changes occurring before major anatomical changes.

I think this "million-dollar question"—i.e. whether evolution driven by behavioral choices/shifts is sufficient to lead to changes of bodyplans and/or the origin of new higher clades—precisely requires a more multidisciplinary approach, such as the one that I hopefully provided in the present book. After reflecting about this for a long time, I now consider that behavior often drives gradual evolutionary changes such as those usually seen in microevolution as well as in major, long-term macroevolutionary trends as the ones illustrated by whales, birds, human bipedalism, or the titanothere horns (Fig. 4.1). In contrast, ONCE can also account for saltational evolution—by way of, e.g. Gerhart and Kirschner's facilitated variation, Waddington's homeorhesis, and/or Goldschmidt's "regulation"—,which probably does occur often in macroevolution and particularly within the rise of new phyla and bodyplans, and in which change of form probably occurred before change of behavior as explained previously. As also noted above, the morphological comparison of for example insect bodies, indicates that related, but different, major clades are associated with the occurrence of homeotic ("macromutational" *sensu* Goldschmidt) changes such as legs being transformed into wings, or a change in the position of the body where legs versus wings are formed (see Figs. 8.3 and 8.4 and Chap. 8). However, it is important to also stress that although in such cases of homeotic changes and saltational evolution behavioral changes are seemingly not the main drivers, they still play a major role in the whole process, which is in contrast to what has been assumed in discussions about "hopeful monsters".

Therefore, a crucial point that must be made is that each of the phenomena mentioned in the summary provided in the beginning of this chapter and shown in Fig. 1.2 can be flexible within the context of ONCE. Surely, there are cases in which random mutations occur and give rise, merely by chance, to phenotypic changes that are selected by natural selection before, or even without, major behavioral changes as postulated by Neo-Darwinists. And, as noted just previously, ONCE also recognizes that many macroevolutionary innovations may have resulted from a major genetic/epigenetic disruption of normal development, thus leading to saltational evolution and/or "hopeful monsters". The take-home message of this book is simply that, at least in cases of long-term directional evolution and macroevolutionary trends—including those leading to many evolutionary innovations as well as in cases leading to evolutionary eco-morphological mismatches—the "behavior/function before form" order of events is probably the most common one. Above all, ONCE's main tenet is that organisms themselves, and especially their behavioral choices/ shifts and persistence, are often the key active players in evolution (Fig. 1.2). However, because life is so rich and complex, it cannot be reduced to a few mechanistic laws and factors that if measured appropriately in detail, would allow us to predict exactly how the behavior or the evolution of a group of organisms will proceed at a certain point in time and space. That is, there is no mechanistic "theory of everything" in Evolutionary Biology, and that is precisely the beauty of it. In this respect, ONCE also integrates ideas from the new *Systems View of Life* described in Capra and Luisi's [50] book. By taking into account organisms and their behaviors, and including information from various different fields and from authors with different opinions, ONCE aims to improve the "understanding of phenomena within the (non-reductionist) context of a larger whole" [50].

This way of thinking is markedly different from a mechanistic/more analytic line of reasoning because one must first understand the whole in order to understand the parts. For instance, emergent proprieties, such as behavioral responses, do not exist at lower levels of complexity, e.g. at the level of atoms (the atoms that form our body cannot walk, nor can they choose to walk, bipedally). As noted by Capra and Luisi, in such a systems view of life, in non-linear, complex systems small changes such as behavioral shifts can cause huge differences. In turn, such points of divergence can theoretically be associated with occasions when unstable parts of complex systems and ancestral order can be broken and thus new order appear. Therefore, such phenomena can also be related to the current Systems Biology discussions on order versus disorder/chaos in evolution and to the possibility that both order and disorder can act simultaneously. This can relate, for example,

Alberch's "Logic of Monsters" to Waddington's homeorhesis and Goldschmidt's "hopeful monsters", as well as the notions of evolutionary innovation and stable-versus-unstable networks to the notion of randomness that is so dear to Gould. In fact, the research performed in my own laboratory is tending increasingly more to focus on studies and discussions related to such subjects, including the use of network theory to undertake anatomical network analyses, and to discuss our results within the context of these broader topics (Chap. 8). This is one of the factors that led to the writing of the present book.

However, as nicely put by Reid [304: 28], the subjects discussed here "are matters for future history, but precisely the name for a synthesis with pan- (such as Matsuda's pan-environmentalism) should remind us to leave the door open for new additions." This is exactly the case of ONCE, which can surely—and must—be expanded, when for instance new empirical data/theoretical ideas will become available. In this work I have simply tried, to the best of my ability, to provide a short discussion of how one can integrate various diverse phenomena and ideas and a systems view of life into a relatively coherent vision of evolution in which organisms are active, and not merely passive, players. I surely have failed to refer to other important phenomena and will probably be accused of discussing/presenting the ones I refer to—or their historical background—in a very simplistic, superficial way. This is the price for my (in this case conscious) behavioral choice to keep this text relatively small and avoid jargon, too-many technical terms, and historical details that could make this book less appealing to the broader public or cause its main message to be somewhat "lost in translation".

Lastly, I would like to say that a main goal of ONCE, and of the selected empirical examples reviewed in this book, is that by emphasizing organisms as key active evolutionary players that are capable of and have the drive to make behavioral choices/shifts, this liberates them from the mechanistic Neo-Darwinist obsession with survival and reproduction. That is, apart from seeing organisms as merely passive in evolution, the view that the major force in evolution is simply related to the differential survival and reproduction of individuals led many evolutionary biologists—in particular ethologists, organismal behaviorists, and many evolutionary psychologists—to defend a very extreme adaptationist view. For instance, within such a view and the typical circular reasoning of the so-called "just-so stories" of adaptationism a wild dog chasing its tail would necessarily be interpreted as the triggering of a predator instinct, and so on. According to such a view, there is seemingly no time for play, or for boredom, or for altruism, or simply for doing something for its own sake. Instead, each and every behavioral trait and action ultimately would have to lead to an increasing ability to eat and/or not be eaten as well as to reproduce.

Unfortunately, as an illustrative example of the problems and implications of such a "struggling" view of evolution in which only the winners can survive, this idea has been overemphasized not only in evolutionary biology and ecology but also in sociological, economical, and philosophical sciences for far too long. Tragic events such as the rise of eugenics and Nazism were in great part only possible because of the disproportionate use, by some people, of such extreme ideas to justify their motivations, thus leading to the horrible consequences that we all know about. Moreover, the view that non-human organisms are merely passive, and have no drive at all, continues to lead to cruel methods employed in numerous behavioral laboratories, as emphasized by De Wall [75]. For instance, many such laboratories keep their animals at 85% of typical body weight to "ensure motivation". Even the famous Yerkes Primate Center went through an early period in which they tested food deprivation on chimpanzees, mainly because behaviorists claimed that this was the only way to give the apes "purpose in life" [75].

Because of the extreme emphasis on this "evolutionary struggle", other phenomena such as mutualistic interactions and the rise of social organization were neglected for far too long. As an example, when the first cases of apparent altruism and cooperation in non-human organisms became known, Neo-Darwinists immediately argued that such cases were based on kinship. This idea has since been disproven in taxa such as primates. For instance, field workers analyzed DNA from feces of chimpanzees and concluded that the vast majority of mutual aid in the forest occurs between unrelated individuals; in fact, there are now various well-documented cases of altruism and cooperation even between members of different species [75]. Fortunately, as noted by Sultan [350: 123–124], "after several decades of focusing largely on antagonistic and competitive effects of species on each other—'conflict and privation' as the major structuring principles for biological communities - facilitation (by positive interactors) is being increasingly integrated into an understanding of community dynamics." For example, "a field experiment across a global sample of montane sites found that interactions among plant species were predominantly positive rather than competitive at high elevations: survival, biomass, and reproductive output of alpine plants were higher when nearby neighbords of other species were present." Numerous cases of symbiosis and cooperation are given in Gilbert and Epel's [151] book *Ecological Developmental Biology*.

Phenomena such as positive interactions, including altruism, are in fact very frequent and important, and in some cases even the most crucial, in the evolution of several taxa, including our own species. There is much more than reproduction, competition, and survival in the fascinating array of behaviors displayed by organisms. There is play, boredom, preference, cooperation, altruism, and, importantly, just doing things for the sake of doing them, particularly in organisms with high cognitive capacities and thus a more diverse array of neurobiological possibilities. When I choose to eat strawberry pie instead of blueberry pie, or my dog sometimes prefers to sleep below the bed while sometimes she sleeps in the bed, these are simply behavioral choices with no final "purpose" or "design" or major implications for our survival and reproduction. Neither me or any part of my body is doing complex mathematical equations to calculate which of the two options—eating blueberry versus strawberry—will enhance the number of babies I will have and the probability that my "genes" will be evolutionarily successful. Surely, none of these actions are neither necessarily a do-or-die moment: they are simply some of the behavioral preferences among the numerous ones that could be held. In this sense, every single second of life is not necessarily a constant struggle for each and every organism inhabiting this planet, where only optimality can exist: life is much more diverse, exciting and fascinating than adaptationists have tried to make us believe.

Chapter 10
General Remarks

(1) As Hoffmeyer and Kull [188: 269–270] profoundly and clearly put it, "organisms do not passively succumb to the severity of environmental judgment; instead, they perceive, interpret, and act in the environment in ways that creatively and unpredictably change the whole setting for selection and evolution."

(2) However, organisms are also constrained by factors intrinsic to themselves, i.e. internal factors—such as developmental internal constraints and internal selection—as well as by factors such as their own behavioral persistence, by natural selection, by genetic drift, and by other phenomena that might lead to a decrease of plasticity. These constraints contribute, on the one hand, toward directing their evolutionary changes—including long-term macroevolutionary trends—but on the other hand they can also lead to eco-morphological mismatches, evolutionary dead-ends, and eventually to their own extinction.

(3) A major point of ONCE that clearly contrasts with Neo-Darwinism is the importance given to organisms themselves, not only due to Baldwin's organic selection in which organisms help to build a niche associated with a specific life style due to their initial behavioral choices and behavioral persistence, but also to constraining internal factors limiting the occupation of the morphospace, as well as to internal selection. It is due to this crucial role played by organisms themselves—in contrast to being merely the passive subject of external forces as defended by externalists—that we do so often see eco-etho-morphological mismatches. This is true not only for the millions of taxa that have become extinct—which clearly did not receive all the attention they deserved in Neo-Darwinism, as they do not do in current Evo-Devo—but even for those taxa that were successful enough to survive until the present day. That is, the more evolutionarily active organisms are, the more one expects to see mismatches. If everything merely concerned random mutations selected by natural selection via current external environmental forces, then one would not expect to see so many mismatches. ONCE lies between externalism and internalism, recognizing the crucial importance of both the external environment (e.g., in natural selection) and internal factors

© Springer International Publishing AG 2017
R. Diogo, *Evolution Driven by Organismal Behavior*,
DOI 10.1007/978-3-319-47581-3_10

(e.g. internal constraints and selection) and in particular the complex, continuous interplay between them. For instance, organismal behavior comes from the organisms themselves but is also often a response to changes in/pressures from the external environment, and many developmental epigenetic phenomena are directly influenced by the environment as well.

(4) As explained throughout the book, ONCE also combines other ideas that have been long considered to be contradictory, such as those defended by Cuvier and by Étienne Geoffroy Saint-Hilaire. In this sense, ONCE is more similar to broader views originally defended by Aristotle and Darwin, who combined for example functionalist and structuralist ideas. The unifying vision of ONCE combines for instance the more mechanistic/materialistic ideas of Lamarck with the holism of authors such as Aristotle and Cuvier, in the sense that one can address mechanisms that are not related to teleological notions without having to recur to a reductionist view of nature and evolution. For example, ONCE recognizes that biological organisms are not just the sum of atoms—as has been historically argued by some reductionists and materialists —but at the same time also defends that organisms are not driven by vitalistic forces, nor designed by supernatural beings for a certain "purpose". That is why ONCE has deep links to systems biology, and in particular to the idea that there are in reality natural explanations for the fact that organisms, and especially their behavior, are much more than just the sum of their atoms or cells. In summary, the aim of ONCE is to provide an holistic explanation for the patterns seen in living organisms by combining ideas of various authors and from diverse schools of thought and avoiding to fall in the trap of having to recur to vitalistic, teleological and religious explanations.

(5) Studies performed in the last decades are increasingly contradicting the gene-centered Neo-Darwinist view of evolution, and the rise of epigenetic and systems biology studies is in turn giving rise to a more holistic framework, which is also expanding/related to a change of frameworks in other fields of science and to broader philosophical aspects concerning our human society. One crucial aspect that is particularly important for ONCE is that behavioral choices affect not only future generations through behavioral persistence associated with natural selection, but also the natural selection of the individuals themselves, through very rapid physiological and anatomical changes within their bodies. That is, in a way the old divide between soul and body that unfortunately continues to plague—in deep, often unconscious ways—not only the broader public and media but also many scientists, is fortunately starting to be increasingly abandoned. I am not an adept of the rising obsession with "body–mind" techniques and therapies because many of them go too far and basically become pseudoscience, thus losing the focus on the scientific data that do show the connection between the body and mind to be much stronger than was assumed in the last century. Such data come for

instance from works with non-primates revealing that behavioral traits related to e.g. social hierarchy can directly influence physiological and even anatomical features within their bodies (e.g. the form of some cells, which in turn affect their health: [321, 345]. Similar recent studies in humans have made a great sensation in the media, revealing for instance that even social behaviors that do not have to do with direct ingestion/inhalation of products such as drugs, tobacco, and alcohol—e.g. getting married, having more friends, and so on— can have a very quick impact in physiological and also anatomical traits (e.g. change in the size of telomeres that changes significantly the chance of dying from cancer: see, e.g., [285]). That is, the so-called "mind-related" traits have a profound impact in the body and thus will affect our own survival and thus also directly influence Darwinian natural (external) selection. For instance, dominant monkeys have a stronger immune system and therefore are more able to resist "attacks" by external microorganisms. In a way, one can say that through their behavioral choices, organisms contribute to "construct" not only their niches during many generations (niche construction) but also their own natural selection ("selection construction") within a single generation.

(6) What are the predictions of ONCE, and how can they be tested? A large set of specific predictions can be extracted from Fig. 1.2, in the sense that it provides a very explicit list of phenomena that are expected to happen along with their links. For instance, it can be predicted that behavioral persistence/inheritance often occurs well outside the range of animals with "culture" and/or with particularly complex neurological systems. It also predicts that hybridization is far more common in and important for evolution then often predicted due to the key role played by organisms and their behaviors. For instance, due to the recognition of a remarkably large number of possible behavioral choices, including random/unexpected ones, such as a bird adopting the song of members of other species or creating new songs based on a mix of those songs and the songs of its parents, which can then potentially lead to hybridization and eventually to the formation of new species. It also predicts that new studies will show how "positive" developmental internal constraints are much more common in evolution than is often assumed. It further foresees that future studies, both on microevolution and macroevolution, will show that randomness has played, and continues to play, a crucial role in the history of life, and that "hopeful monsters" were and are more important for evolution than normally predicted, in particular with respect to major morphological innovations. It is also expected that developmental empirical studies will show the importance of facilitated variation and homeorhesis for not only the rise and viability of such "hopeful monsters" but also for a much broader range of evolutionary outcomes. Importantly, ONCE predicts that empirical ecomorphological studies using a

rigorous phylogenetic framework will continue to reveal that cases of eco-morphological mismatches are much more common than has been assumed by Darwinists, Neo-Darwinists, and even by current Evo-Devoists, including the proponents of Extended Evolutionary Synthesis.

(7)   In addition to predicting the existence/importance of such phenomena, ONCE also predicts that in general—but not always—these phenomena will follow a certain temporal order, as shown in Fig. 1.2. For instance, against Neo-Darwinism, it envisages that behavior/function normally changes before form because organisms and their behavioral shifts/persistence are the primary, active players in evolution, and natural selection is mainly a secondary main player. However, ONCE does also predict that, as an exception to this rule, in many cases—particularly in major morphological changes/innovations such as those related to the occurrence of "hopeful monsters"—form can change before function/behavior. The occurrence of such cases is more likely within the more relaxed, nonoptimal, "nonstruggling" view of evolution of ONCE. That is, as long as "hopeful monsters" are not directly and strongly detrimental, they can survive in a non-optimal world; later, their new phenotypes can open up a wide range of new behavioral choices/life in new niches, as discussed in Chap. 8.

(8)   Darwin did care a lot about extinction, and was much more aware of evolutionary mismatches than were most Neo-Darwinists. For instance, extinction is included in the last, and most famous, sentences of his 1959 book: "*A struggle for life, and as a consequence to natural selection, entailing divergence of character and the extinction of less-improved forms... thus, from the war of nature, from famine and death, the most exalted object which we are capable of conceiving, namely, the production of the higher animals, directly follows... there is grandeur in this view of life, with its several powers, having been originally breathed into a few forms or into one; and that, whilst this planet has gone cycling on according to the fixed law of gravity, from so simple a beginning endless forms most beautiful and most wonderful have been, and are being, evolved*". It is interesting to note that many of the points in ONCE—and that contradict the ideas of many, if not most, Neo-Darwinists—were made previously by Darwin, as explained throughout this book (see, e.g., the part about plants in Chap. 3). In fact, the central tenet defended in this book and in ONCE—that organismal behavioral is the central, key evolutionary player—was defended in Darwin's early writings: in his notebooks M and N Darwin asks how is "good design" achieved and then answered to himself with a single word: "behavior" [19: p. 115].

(9)   Regarding the last sentences of his 1859 book cited previously, I also completely agree with Darwin's emphasis on extinction and "less-improved

forms" to refer to evolutionary mismatches, and in some ways some of the tenets of the idea of ONCE are also not too different from those of Aristotle, who is one of my personal heroes. However, unlike Darwin I would not use the teleological word "improved" in this context. Above all, I think that the new discoveries in evolutionary biology that were only possible because of Darwin have showed us that life is not only an intense, never ending, unbreakable "struggle, war, famine and death", in which forms must optimally fit themselves to their habitats and that any non-optimality, any behavior that is not directly related to survival and/or reproduction and does not improve fitness, is purged from existence.

(10) Interestingly, the notion of "struggle" for existence and of optimality dates far back in time, including to Aristotle, who famously stated that nature "does nothing in vain" [223]. For instance, a recent book that I otherwise liked very much is a clear example of this still-dominant conception of life, often providing statements such as: "in the plant world as in the animals, no one does anything for nothing" [241: 110; see chap. 3]. However, as noted in the Preface, more and more studies are emphasizing the large plasticity between the so-called "optimal" morphology of a structure and the potential function of that structure, underscoring the need to appreciate apparently "maladaptive" structures in biological evolution as nevertheless effective functioning units. That is, such structures and the function they perform are "good enough" to allow the organisms displaying them to survive and reproduce, in the nonstruggling view of life defended in ONCE. Not every single structure needs to be "perfectly designed", or "optimal", or "optimally fit" for each function it performs or for each habitat occupied or behavior exhibited by the organism displaying it, at every single time during its evolutionary history: in fact, many structures, such as vestigial rudimentary structures, may well not perform any function at all.

(11) I hope that I have convinced the readers that, fortunately, life on earth is much more diverse, complex, unpredictable, and therefore fascinating than it is suggested under the notion of "struggle" for existence and of optimality. As long as overspecialized humans do not put in danger the ecosystems of this beautiful planet, and the globe continues to display such an amazing quality and diversity of resources, life it much richer and varied than that. For instance, analyses of the yeast genome have shown that 70% of its genes were actually unnecessary in a rich medium. If this is really so, even if we follow a "utilitarian" framework and speculate that all those 70% may eventually become useful for subsequent particular environmental changes [327], the fact is that when yeast lives in such a rich medium, more than two thirds of their genes are "dispensable" or at least redundant. This does not match at all with the narrative idea of a continuous, implacable struggle in evolution, in which "nature does nothing in vain". This planet has seen, since

approximately 4 billion years ago, zillions of fascinating organisms displaying an almost infinite array of behaviors from bacteria's adaptive comportments to suicidal or self-harming acts by birds, from selfish attitudes of cuckoos to profoundly altruistic warning calls of squirrels, from remarkable phenotypic features to expandable genes. It is this fascinating mix between an incredible diversity of behaviors, of forms adapted to certain habitats, of non-optimal morphologies, of eco-etho-morphological mismatches, and of a high proportion of random events, what has precisely led to Darwin's *"endless forms most beautiful and most wonderful"*.

References

1. Abdala V, Tulli MJ, Russell AP, Powell GL, Cruz FB. 2014. Anatomy of the crus and pes of Neotropical iguanian lizards in relation to habitat use and digitally based grasping capabilities. Anat Rec 297:397–409.
2. Abzhanov A. 2013. Von Baer's law for the ages: lost and found principles of developmental evolution. Trends Genet 29:712–722.
3. Ackermann RR, Mackay A, Arnold ML. 2016. The hybrid origin of "modern" humans. Evol Biol 43:1–11.
4. Adamowicz SJ, Purvis A, Wills MA. 2008. Increasing morphological complexity in multiple parallel lineages of the Crustacea. PNAS 105:4786–4791.
5. Albalat R, Canestro C. 2016. Evolution by gene loss. Nat Rev Genet: 10.1038/nrg.2016.39.
6. Alberch P. 1985. Developmental constraints: why St. Bernards often have an extra digit and poodles never do. Am Naturalist 126:430–433.
7. Alberch P. 1989. The logic of monsters: evidence for internal constraint in development and evolution. Geobios Mém Spéc 12:21–57.
8. Alberch P, Gale EA. 1983. Size dependence during the development of the amphibian foot - Colchicine-induced digital loss and reduction. J Embryol Exp Morphol 76:177–197.
9. Alberch P, Gale EA. 1985. A developmental analysis of an evolutionary trend: digital reduction in amphibians. Evolution 39:8–23.
10. Allen JA. 1877. The influence of physical conditions in the genesis of species. Radical Rev 1:108–140.
11. Alroy J. 2000. Understanding the dynamics of trends within evolving lineages. Paleobiology 26:707–733.
12. Almécija S, Smaers JB, Jungers WL. 2015. The evolution of human and ape hand proportion. Nature Communications 6:7717.
13. Ancet P. 2006. L'observation des monstres dans l'œuvre d'Etienne Geoffroy Saint-Hilaire. Cah Philos 108:23–38.
14. Anderson TM, von Holdt BM, Candille SI, Musiani M, Greco C, Stahler DR, Barsh GS. 2009. Molecular and evolutionary history of melanism in North American gray wolves. Science 323:1339–1343.
15. Arthur W. 1997. The origin of animal body plans: a study in evolutionary developmental biology. Cambridge University Press, Cambridge.
16. Arthur W. 2002. The emerging conceptual framework of evolutionary developmental biology. Nature 415:757–764.
17. Arthur W. 2004. Biased embryos and evolution. Cambridge University Press, Cambridge.
18. Arthur W. 2011. Evolution - a developmental approach. Wiley-Blackwell, Oxford.
19. Asma ST. 1996. Following form and function - a philosophical archaeology of life science. Northwestern University Press, Evanston.
20. Aversi-Ferreira TA, Maior RS, Carneiro-e-Silva FO, Aversi-Ferreira RAG, Tavares MC, Nishijo H, Tomaz C. 2011. Comparative anatomical analyses of the forearm muscles of Cebus libidinosus (Rylands et al. 2000): manipulatory behavior and tool use. Plos One 6:1–8.
21. Baab KL, Perry JMG, Rohlf FJ, Jungers WL. 2014. Phylogenetic, ecological, and allometric correlates of cranial shape in Malagasy lemuriforms. Evolution 68:1450–1468.
22. Badyaev AV. 2011. Origin of the fittest: link between emergent variation and evolutionary change as a critical question in evolutionary biology. Proc R Soc Biol Sci B 278:1921–1929.

© Springer International Publishing AG 2017
R. Diogo, *Evolution Driven by Organismal Behavior*,
DOI 10.1007/978-3-319-47581-3

23. Baldwin JM. 1895. Mental development in the child and race: methods and processes. MacMillan, New York.

24. Baldwin JM. 1896a. A new factor in evolution. Am Naturalist 30:441–451.

25. Baldwin JM. 1896b. A new factor in evolution (continued). Am Naturalist 30:536–553.

26. Baldwin JM. 1896c. On criticisms of organic selection. Science 4:724–727.

27. Balon EK. 2004. Alternative ontogenies and evolution: a farewell to gradualism. In Environment, Development and Evolution: Toward a Synthesis (The Vienna Series in Theoretical Biology) (Hall BK, Pearson RD, Müller GB, eds.). A Bradford Book, Massachusetts. p. 37–66.

28. Barash BA, Freedman L, Opitz JM. 1970. Anatomical studies in the 18-trisomy syndrome. Birth Defects 6:3–15.

29. Bardeen CR. 1906. Development and variation of the nerves and the musculature of the inferior extremity and of the neighboring regions of the trunk in man. Am J Anat 6:259–390.

30. Bardeen CR. 1910. Development of the skeleton and of the connective tissues. Manual Human Embryol 1:438–439.

31. Barron AB, Klein C. 2016. What insects can tell us about the origins of consciousness? PNAS 113:4900–4908.

32. Bartels M. 1880-1881. Ueber Menschenschwänze. Archiv für Anthropol 13:1–41.

33. Bateson P, Gluckman P, Hanson M. 2014. The biology of developmental plasticity and the predictive adaptive response hypothesis. J Physiol 592: 2357–2368.

34. Bateson W. 1894. Materials for the study of variation treated with especial regard to discontinuity in the origin of species. Macmillan, London.

35. Beatty J. 2008. Chance variation and evolutionary contingency: Darwin, Simpson (The Simpsons) and Gould. In The Oxford handbook of philosophy of biology (Ruse M, ed.). Oxford University Press, Oxford. p. 189–210.

36. Bhullar B-AS, Morris ZS, Sefton EM, Tok A, Tokita M, Namkoong B, Camacho J, Burnham DA, Abzhanov A. 2015. A molecular mechanism for the origin of a key evolutionary innovation, the bird beak and palate, revealed by an integrative approach to major transitions in vertebrate history. Evolution 69:1665–1677.

37. Bininda-Emonds OR, Jeffery JE, Richardson MK. 2003. Inverting the hourglass: quantitative evidence against the phylotypic stage in vertebrate development. Proc Biol Sci 270:341–346.

38. Birkhead TR, Monaghan P. 2010. Ingenious ideas: a history of behavioral ecology. In: Evolutionary Behavioral Ecology (Westneat DF, Fox CW, eds.). Oxford University Press, New York. p. 3–15.

39. Blackiston D, Silva Casey E, Weiss M. 2008. Retention of memory through metamorphosis: can a moth remember what it learned as a caterpillar? PLOS One 3:e1736.

40. Blake RW, Chan KHS. 2007. Swimming in the upside down catfish Synodontis nigriventris: it matters which way is up. J Exp Biol 210: 2979–2989.

41. Blumberg MS. 2009. Freaks of nature: what anomalies tell us about development and evolution. Oxford University Press, New York.

42. Bocherens H, Schrenk F, Chaimanee Y, Kullmer O, Mörike D, Pushkina D, Jaeger J-J. 2016. Flexibility of diet and habitat in Pleistocene South Asian mammals: implication for the fate of the giant fossil ape Gigantopithecus. Quaternary Internat. In press.

43. Bolk L. 1926. Das Problem der Menschwerdung. Fischer, Jena.

44. Bonner JT. 1988. The evolution of complexity by means of natural selection. Princeton University Press, Princeton.

45. Bonner JT. 2013. Randomness in Evolution. Princeton University Press, Princeton.

46. Brigandt I. 2003. Homology and the origin of correspondence. Biol & Philos 17:389–407.

47. Brooks RC, Griffith SC. 2010. Mate choice. In Evolutionary behavioral ecology (Westneat DF, Fox CW, eds.). Oxford University Press. New York. p. 416–433.

48. Bufill E, Agusti J, Blesam R. 2011. Human neoteny revisited: the case of synaptic plasticity. Am J Hum Biol 23:729–739.
49. Butler FP. 2012. Evolution without Darwinism - the legacy of Stephen Jay Gould. CreateSpace, New York.
50. Capra F, Luisi PL. 2014. The systems view of life. Cambridge University Press, Cambridge.
51. Cardini A, Elton S. 2008. Variation in guenon skulls (I): species divergence, ecological and genetic differences. J Hum Evol 54:615–637.
52. Carrizo LV, Tulli MJ, Abdala V. 2014. An ecomorphological analysis of forelimb musculotendinous system in sigmodontine rodents (Rodentia, Cricetidae, Sigmodontinae). J Mammal 95:843–854.
53. Carroll RL. 1997. Patterns and processes of vertebrate evolution. Cambridge University Press, Cambridge.
54. Carroll SB. 2016. The Serengeti rules: the quest to discover how life works and why it matters. Princeton University Press, Princeton.
55. Carroll SB, Weatherbee S, Langeland J. 1995. Homeotic genes and the regulation and evolution of insect wing number. Nature 375:58–61.
56. Chemisquy MA, Prevosti FJ, Martin G, Flores DA. 2015. Evolution of molar shape in didelphid marsupials (Marsupialia: Didelphidae): analysis of the influence of ecological factors and phylogenetic legacy. Zool J Linn Soc 173:217–235.
57. Cihak R. 1972. Ontogenesis of the skeleton and intrinsic muscles of the human hand and foot. Adv Anat Embryol Cell Biol 46:1–194.
58. Clemente CJ. 2014. The evolution of bipedal running in lizards suggests a consequential origin may be exploited in later lineages. Evolution 68:2171–2183.
59. Collar D, Wainwright P. 2006. Discordance between morphological and mechanical diversity in the feeding mechanism of centrarchid fishes. Evolution 60:2575–2584.
60. Cope ED. 1896. The primary factors of organic evolution. Open Court Publishing Co., Chicago.
61. Corning PA. 2013. Evolution 'on purpose': how behavior has shaped the evolutionary process. Biol J Linn Soc 112:242–260.
62. Cossetti C, Lugini L, Astrologo L, Saggio I, Fais S, Spadafora C. 2014. Soma-to-germline transmission of RNA in mice xenografted with human tumour cells: possible transport by exosomes. PLoS ONE 9:e101629.
63. Coulson T, MacNulty DR, Stahler DR, von Holdt B, Wayne RK, Smith DW. 2011. Modeling effects of environmental change on wolf population dynamics, trait evolution, and life history. Science 334:1275–1278.
64. Crispo E. 2007. The Baldwin effect and genetic assimilation: revisiting two mechanisms of evolutionary change mediated by phenotypic plasticity. Evolution 61:2469–2479.
65. Cruz FB, Belver L, Acosta JC, Villavicencio HJ, Blanco G, Canovas MG. 2009. Thermal biology of Phymaturus lizards: evolutionary constraints or lacks of environmental variation. Zoology 112:425–432.
66. Danowitz M, Vasilyev A, Kortlandt V, Solounias V. 2015. Fossil evidence and stages of elongation of the Giraffa camelopardalis neck. R Soc Open Sci 2:150393.
67. Darwin C. 1859. On the origin of species by means of natural selection, or, the preservation of favored races in the struggle for life. J. Murray, London.
68. Darwin C, Darwin F, 1880. The power of movement in plants. John Murray, London.
69. Davidson E. 2006. The regulatory genome: gene regulatory networks in development and evolution. Academic Press, San Diego.
70. Davidson PE. 1914. The recapitulation theory and human infancy. Teachers College, New York.
71. Dawkins R. 1982. The extended phenotype. WH Freeman & Co, Oxford.
72. De Beer G. 1940. Embryos and ancestors. Clarendon Press, Oxford.

73. Dececchi et al. (2016), The wings before the bird: an evaluation of flapping-based locomotory hypotheses in bird antecedents. PeerJ 4:e2159.

74. De Panafieu J-B. 2007. Evolution. Seven Stories Press, New York.

75. De Wall F. 2016. Are We Smart Enough To Know How smarts Animal Are? WWW Norton & Company Inc, New York.

76. Depew DJ. 2003. Baldwin and his many effects. In Evolution and learning: The Baldwin effect reconsidered (Weber BH, Depew DJ, eds.). MIT Press, Cambridge. p. 3–31.

77. Dial KP, Jackson BE, Segre P. 2008. A fundamental avian wing-stroke provides a new perspective on the evolution of flight. Nature 451:985–989.

78. Dial KP, Heers AM, Dial TR. 2015. Ontogenetic and evolutionary transformations in avian locomotion: the ecological significance of rudimentary structures. In Great transformations: Major Events in the History of Vertebrate Life (Dial KP, Shubin N, Brainerd EL, eds.). University of California Press, Berkeley. p. 283–301.

79. Diaz RE, Trainor PA. 2015. Hand/foot splitting and the 're-evolution' of mesopodial skeletal elements during the evolution and radiation of chameleons. BMC Evol Biol 15:184.

80. Diogo R. 2004. Morphological evolution, aptations, homoplasies, constraints, and evolutionary trends: catfishes as a case study on general phylogeny and macroevolution. Science Publishers, Enfield.

81. Diogo R. 2006. Cordelia's dilemma, historical bias, and general evolutionary trends: catfishes as a case study for general discussions on phylogeny and macroevolution. Internat J Morphol 24:607–618.

82. Diogo R. 2007. On the origin and evolution of higher-clades: osteology, myology, phylogeny and macroevolution of bony fishes and the rise of tetrapods. Science Publishers, Enfield.

83. Diogo R. 2010. Comparative anatomy, anthropology and archaeology as case studies on the influence of human biases in natural sciences: the origin of 'humans', of 'behaviorally modern humans' and of 'fully civilized humans'. Open Anat J 2:86–97.

84. Diogo R. 2016. Where is the Evo in Evo-Devo (Evolutionary Developmental Biology). J Exp Zool B (Mol Dev Evol) 326:9–18.

85. Diogo R, Abdala V. 2010. Muscles of vertebrates – comparative anatomy, evolution, homologies and development. Taylor & Francis, Oxford.

86. Diogo R, Wood B. 2011. Soft-tissue anatomy of the primates: phylogenetic analyses based on the muscles of the head, neck, pectoral region and upper limb, with notes on the evolution of these muscles. J Anat 219:273–359.

87. Diogo R, Tanaka EM. 2012. Anatomy of the pectoral and forelimb muscles of wildtype and green fluorescent protein-transgenic axolotls and comparison with other tetrapods including humans: a basis for regenerative, evolutionary and developmental studies. J Anat 221:622–635.

88. Diogo R, Wood B. 2012a. Comparative anatomy and phylogeny of primate muscles and human evolution. Taylor and Francis, Oxford.

89. Diogo R, Wood B. 2012b. Violation of Dollo's Law: evidence of muscle reversions in primate phylogeny and their implications for the understanding of the ontogeny, evolution and anatomical variations of modern humans. Evolution 66:3267–3276.

90. Diogo R, Wood B. 2013. The broader evolutionary lessons to be learned from a comparative and phylogenetic analysis of primate muscle morphology. Biol Rev 88:988–1001.

91. Diogo R, Molnar JL. 2014. Comparative anatomy, evolution and homologies of the tetrapod hindlimb muscles, comparisons with forelimb muscles, and deconstruction of the forelimb-hindlimb serial homology hypothesis. Anat Rec 297:1047–1075.

92. Diogo R, Tanaka EM. 2014. Development of fore- and hindlimb muscles in GFP-transgenic axolotls: Morphogenesis, the tetrapod Bauplan, and new insights on the forelimb-hindlimb enigma. J Exp Biol B (Mol Dev Evol) 322: 106–127.

93. Diogo R, Ziermann JM. 2014. Development of fore- and hindlimb muscles in frogs: Morphogenesis, homeotic transformations, digit reduction, and the forelimb-hindlimb enigma. J Exp Biol B (Mol Dev Evol) 322:86–105.
94. Diogo R, Ziermann JM. 2015a. Development, metamorphosis, morphology and diversity: the evolution of chordate muscles and the origin of vertebrates. Dev Dyn.:10.1002/dvdy.24245.
95. Diogo R, Ziermann JM. 2015b. Muscles of chondrichhtyan paired appendages: comparison with osteichthyans, deconstruction of the fore-hindlimb serial homology dogma, and new insights on the evolution of the vertebrate neck. Anat Rec 298:513–530.
96. Diogo R, Wood B. 2016. Origin, development and evolution of primate muscles, with notes on human anatomical variations and anomalies. In Developmental approaches to human evolution (Boughner J, Rolian C, eds.). John Wiley & Sons, Hoboken. p. 167–204.
97. Diogo R, Molnar JL. 2016. Links between evolution, development, human anatomy, pathology and medicine, with a proposition of a re-defined anatomical position and notes on constraints and morphological 'imperfections'. J Exp Zool B (Mol Dev Evol). In press.
98. Diogo R, Santana SE. 2016. Evolution of facial musculature and relationships with facial color patterns, mobility, social group size, development, birth defects and assymetric use of facial expressions. In The Science of Facial Expression (Russel R, Dols JMF, eds.). Oxford University Press, Oxford. In press.
99. Diogo R, Abdala V, Lonergan NL, Wood B. 2008a. From fish to modern humans - comparative anatomy, homologies and evolution of the head and neck musculature. J Anat 213:391–424.
100. Diogo R, Hinits Y, Hughes S. 2008b. Development of mandibular, hyoid and hypobranchial muscles in the zebrafish, with comments on the homologies and evolution of these muscles within bony fish and tetrapods. BMC Dev Biol 8:24–46.
101. Diogo R, Abdala V, Aziz MA, Lonergan NL, Wood B. 2009a. From fish to modern humans - comparative anatomy, homologies and evolution of the pectoral and forelimb musculature. J Anat 214:694–716.
102. Diogo R, Wood B, Aziz MA, Burrows A. 2009b. On the origin, homologies and evolution of primate facial muscles, with a particular focus on hominoids and a suggested unifying nomenclature for the facial muscles of the Mammalia. J Anat 215:300–319.
103. Diogo R, Richmond BG, Wood B. 2012. Evolution and homologies of primate and modern human hand and forearm muscles, with notes on thumb movements and tool use. J Hum Evol 63:64–78.
104. Diogo R, Linde-Medina M, Abdala V, Ashley-Ross MA. 2013a. New, puzzling insights from comparative myological studies on the old and unsolved forelimb/hindlimb enigma. Biol Rev 88:196–214.
105. Diogo R, Peng, Z, Wood B. 2013b. First comparative study of morphological and molecular evolutionary rates within primates: implications for the tempo and mode of primate and human evolution. J Anat 222:410–418.
106. Diogo R, Smith C, Ziermann JM. 2015a. Evolutionary Developmental Pathology and Anthropology: a new area linking development, comparative anatomy, human evolution, morphological variations and defects, and medicine. Dev Dyn 244:1357–1374.
107. Diogo R, Esteve-Altava B, Smith C, Boughner JC, Rasskin-Gutman. 2015b. Anatomical network comparison of human upper and lower, newborn and adult, and normal and abnormal limbs, with notes on development, pathology and limb serial homology vs. homoplasy. PLOS One:10.1371/journal.pone.0140030.
108. Diogo R, Ziermann JM, Linde-Medina M. 2015c. Specialize or risk disappearance - empirical evidence of anisomerism based on comparative and developmental studies of gnathostome head and limb musculature. Biol Rev 90:964–978.
109. Diogo R, Ziermann JM, Linde-Medina M. 2015d. Is evolutionary biology becoming too politically correct? A reflection on the scala naturae, phylogenetically basal clades, anatomically plesiomorphic taxa, and `lower' animals. Biol Rev 90:502–521.

110. Diogo R, Kelly R, Christian L, Levine M, Ziermann JM, Molnar J, Noden D, Tzahor E. 2015e. A new heart for a new head in vertebrate cardiopharyngeal evolution. Nature 520:466–473.

111. Diogo R, Walsh S, Smith C, Ziermann JM, Abdala V. 2015f. Towards the resolution of a long-standing evolutionary question: muscle identity and attachments are mainly related to topological position and not to primordium or homeotic identity of digits. J Anat 226:523–529.

112. Diogo R, Noden D, Smith CM, Molnar JL, Boughner J, Barrocas C, Bruno J. 2016. Learning and understanding human anatomy and pathology: an evolutionary and developmental guide for medical students. Taylor & Francis, Oxford.

113. Diogo R, Guinard G, Diaz R (2017, in press). Dinosaurs, chameleons, humans and Evolutionary Developmental Pathology: linking Étienne Geoffroy St. Hilaire, Waddington's homeorhesis, Alberch's logic of monsters, and Goldschmidt hopeful monsters. J Exp Zool B (Mol Dev Evol).

114. Dittrich-Reed DR, Fitzpatrick BM. 2013. Transgressive hybrids as hopeful monsters. Evol Biol 40:310–315.

115. Downes SM. 2003. Baldwin effects and the expansion of the explanatory repertoire in evolutionary biology. In Evolution and learning: The Baldwin effect reconsidered (Weber BH, Depew DJ, eds.). MIT Press, Cambridge. p. 33–52.

116. Duhamel B. 1972. L'œuvre tératologique d'Etienne Geoffroy Saint-Hilaire. Re Hist Sci 25:337–346.

117. Duckworth RA. 2009. The role of behavior in evolution: a search for mechanisms. Evol Ecol 23:513–531.

118. Dunlap SS, Aziz MA, Rosenbaum KN. 1986. Comparative anatomical analysis of human trisomies 13, 18 and 21: I, the forelimb. Teratology 33:159–186.

119. Eble GJ. 2004. The macroevolution of phenotypic integration. In Phenotypic integration: studying the ecology and evolution of complex phenotypes (Pigliucci M, Preston K, eds.). Oxford, New York. p. 253–272.

120. Eble GJ. 2005. Morphological modularity and macroevolution: conceptual and empirical aspects. In Modularity: understanding the development and evolution of natural complex systems (Callebaut W, Rasskin-Gutman D, eds.). MIT Press Books, Cambridge. p. 221–238.

121. Eldredge N. 2014. Extinction and evolution: what fossils reveal about the history of life. Firefly Books, Toronto

122. Eldredge N, Gould SJ. 1972. Punctuated equilibrium: an alternative to phyletic gradualism. In Models in paleobiology (Schopf TJM, ed.). Freeman, Cooper and Co., San Francisco. p. 82–115.

123. Emmons-Bell M, Durant F, Hammelman J, Bessonov N, Volpert V, Morokuma J, Pinet K, Adams DS, Pietak A, Lobo D, Levin M. 2015. Gap junctional blockade stochastically induces different species-specific head anatomies in genetically wild-type Girardia dorotocephala flatworms. Int J Mol Sci 16:27865–27896.

124. Erwin DH. 2015. Novelty and innovation in the history of life. Curr Biol 25:R930-R940.

125. Esteve-Altava B, Marugán-Lobón J, Botella H, Rasskin-Gutman D. 2013. Structural constraints in the evolution of the tetrapod skull complexity: Williston's law revisited using network models. Evol Biol 40:209–219.

126. Esteve-Altava B, Diogo R, Smith C, Boughner JC, Rasskin-Gutman D. 2015. Anatomical networks reveal the musculoskeletal modularity of the human head. Sci Rep 5:8298.

127. Fabre A-C, Cornette R, Peigné S, Goswami A. 2013. Influence of body mass on the shape of forelimb in musteloid carnivorans. Biol J Linn Soc 110:91–103.

128. Fabre A-C, Cornette R, Goswami A, Peigné S. 2015. Do constraints associated with the locomotor habitat drive the evolution of forelimb shape? A case study in musteloid carnivorans. J Anat 226:596–610.

129. Fabrezi M, Manzano AS, Abdala A, Lobo F. 2014. Anuran locomotion: ontogeny and morphological variation of a distinctive set of muscles. Evol Biol 41:308–326.
130. Falkner G, Falkner R. 2013. On the imcompability of the Neo-Darwinian hypothesis with systems-theoretical explanations of biological development. In Beyond Mechanism: Putting Life Back into Biology (Henning BG, Scarfe AC, eds). Lexington Books, Lexington. p. 93–114.
131. Fisher RA. 1930. The Genetical Theory of Natural Selection. Clarendon Press, Oxford.
132. Fisher RA. 1958. The Genetical Theory of Natural Selection, 2nd Ed. Dover, New York.
133. Fox CW, Westneat DF. 2010. Adaptation. In Evolutionary behavioral ecology (Westneat DF, Fox CW, eds.). Oxford University, New York. p. 16–32.
134. Fricke C, Bretman A, Chapman T. 2010. Sexual conflict. In Evolutionary behavioral ecology (Westneat DF, Fox CW, eds.). Oxford University, New York. p. 400–415.
135. Futuyma DJ. 2010. Evolutionary constraint and ecological consequences. Evolution 64:1865–1884.
136. Futuyma DJ. 2013. Evolution, 3rd Ed. Sinauer Associates, Inc. Publishers. Chicago.
137. Gailer JP, Calandra I, Schulz-Kornas E, Kaiser TM. 2016. J Mammal Evol 23:369–383.
138. Galis F. 1999. Why do almost all mammals have seven cervical vertebrae? Developmental constraints, Hox genes, and cancer. J Exp Zool 285:19–26.
139. Galis F, Metz JAJ. 2007. Evolutionary novelties: the making and breaking of pleiotropic constraints. Integr Compar Biol 47:409–419.
140. Galis F, Terlouw A, Osse JWM. 1994. The relation between morphology and behavior during ontogenic and evolutionary changes. J Fish Biol 45:13–26.
141. Gasser RF. 1967. The development of the facial muscles in man. Am J Ant 120:357–376.
142. Geoffroy Saint-Hilaire É. 1818-1822. Philosophie anatomique. Baillière, Paris.
143. Geoffroy Saint-Hilaire I. 1832-1837. Histoire générale et particulière des anomalies de l'organisation chez l'homme et les animaux. Baillière, Paris.
144. Gerhart J, Kirschner M. 2005. The plausibility of life: resolving Darwin's dilemma. Yale University Press, Yale.
145. Gerhart J, Kirschner M. 2007. The theory of facilitated variation. PNAS 104 Suppl 1:8582–8589.
146. Ghalambor CK, Angeloni LM, Carroll SP. 2010. Behavior as phenotypic plasticity. In Evolutionary behavioral ecology (Westneat DF, Fox CW, eds.). Oxford University Press, New York. p. 90–107.
147. Gilbert SF. 1988. Cellular politics: Ernest Everett Just, Richard B. Goldschmidt, and the attempt to reconcile embryology and genetics. In The American Development of Biology (Rainger R, Benson K, Maienschein J, eds.). University Pennsylvania Press, Philadelphia. p. 311–346.
148. Gilbert SF. 2003. The role of predator-induced polyphenism in the evolution of cognition: a Baldwinian speculation. In Evolution and Learning: The Baldwin Effect Reconsidered (Weber BH, Depew DJ). MIT Press, Cambridge. p. 235–252.
149. Gilbert SF. 2006. Developmental biology, 8th Ed. Sinauer Associates Inc., Sunderland.
150. Gilbert SF. 2010. Developmental biology, 9th Ed. Sinauer Associates Inc., Sunderland.
151. Gilbert SF, Epel D. 2009. Ecological developmental biology - integrating epigenetics, medicine and evolution. Sinauer Associates, Inc., Sunderland.
152. Goldschmidt R. 1940. The material basis of evolution. Yale Univ Press, New Haven.
153. Goodman BA, Miles DB, Schwarzkopf L. 2008. Life on the rocks: habitat use drives morphological and performance evolution in lizards. Ecology 89:3462–3471.
154. Goodwin BC. 1982. Development and evolution. J Theoret Biol 97:43–55.
155. Goodwin BC. 1984. Changing from an evolutionary to a generative paradigm in biology. In Evolutionary theory: paths into the future (Pollard JW, ed.). John Wiley & Sons, Chicester. p. 99–120.
156. Gould SJ. 1977. Ontogeny and phylogeny. Harvard University Press, Cambridge.

157. Gould SJ. 1989. Wonderful life: the Burgess shale and the nature of history. Norton WW, New York.

158. Gould SJ. 1993. Cordelia's dilemma. Nat Hist 102:10–18.

159. Gould SJ. 2002. The structure of evolutionary theory. Belknap, Harvard.

160. Guinard G. 2015. Introduction to evolutionary teratology, with an application to the forelimbs of Tyrannosauridae and Carnotaurinae (Dinosauria: Theropoda). Evol Biol 42:20–41.

161. Griffiths PE. 2003. Beyond the Baldwin effect: James Mark Baldwin's 'social heredity', epigenetic inheritance and niche-construction. In Evolution and Learning: The Baldwin Effect Reconsidered (Weber BH, Depew DJ). MIT Press, Cambridge. p. 193–215.

162. Gühmann M, Jia H, Randel N, Verasztó C, Bezares-Calderón LA, Michiels NK, Yokoyama S, Jékely G. 2015. Spectral tuning of phototaxis by a Go-Opsin in the rhabdomeric eyes of Platynereis. Curr Biol 25:2265–2271.

163. Grant PR, Grant BR. 2009. The secondary contact phase of allopatric speciation in Darwin's finches. PNAS 106:20141–20148.

164. Grant BR, Grant PR. 2010. Songs of Darwin's finches diverge when a new species enters the community. PNAS 107:20156–20163.

165. Grizante MB, Navas CA, Garland T jr, Kohlsdorf T. 2010. Morphological evolution in Tropidurinae squamates: an integrated view along a continuum of ecological settings. J Evol Biol 3:98–111.

166. Hale RE, Miller N, Francis RA, Kennedy C. 2016. Does breeding ecology alter selection on developmental and life history traits? A case study in two ambystomatid salamanders. Evol Ecol 30:503–517.

167. Hall BK. 1984. Developmental mechanisms underlying the formation of atavisms. Biol Rev 59:89–124.

168. Hall BK. 2001. Organic selection: proximate environmental effects on the evolution of morphology and behavior . Biol & Philos 16:215–237.

169. Hall BK. 2002. Palaeontology and Evolutionary Developmental Biology: a science of the nineteenth and twenty–first centuries. Palaeontology 45:647–669.

170. Hall BK. 2003. Baldwin and beyond: organic selection and genetic assimilation. In Evolution and Learning: The Baldwin Effect Reconsidered (Weber BH, Depew DJ). MIT Press, Cambridge. p. 141–167.

171. Hall BK. 2004. Introduction: evolution as the control of development by ecology. In Environment, Development and Evolution: Toward a Synthesis (The Vienna Series in Theoretical Biology) (Hall BK, Pearson RD, Müller GB, eds.). A Bradford Book, Massachusetts. p. ix-xxiii.

172. Hall BK, Pearson RD, Müller GB, eds. (2004) Environment, Development and Evolution: Toward a Synthesis (The Vienna Series in Theoretical Biology). A Bradford Book, Massachusetts.

173. Hansen TF, Houle D. 2004. Evolvability, stabilizing selection, and the problem of stasis. In Phenotypic integration: studying the ecology and evolution of complex phenotypes. (Pigliucci M, Preston K., eds.). Oxford University press, Oxford.

174. Hardy AC. 1965. The living stream - a restatement of evolution theory and its relation to the spirit of man. Collins, London.

175. Hartl DL. 2009. Essential genetics - a genomics perspective. Jones & Bartlett Learnin, New York.

176. Hazkani-Covo E, Wool D, Graur D. 2005. In search of the vertebrate phylotypic stage: a molecular examination of the developmental hourglass model and von Baer's third law. J Exp Zool (Mol Dev Evol) 304B: 150–158.

177. Healy SD, Rowe C . 2010. Information processing: the ecology and evolution of cognitive abilities. In Evolution behavioral ecology (Westneat DF, Fox CW, eds.). Oxford University Press, New York. p. 162–174

178. Heers AM, Baier DB, Jackson BE, Dial KP. 2016. Flapping before flight: high resolution, three-dimensional skeletal kinematics of wings and legs during avian development. PLOS One 11:e0153446.
179. Heiss E, Handschuh S, Aerts P, Van Wassenbergh S. 2016. Musculoskeletal architecture of the prey capture apparatus in salamandrid newts with multiphasic lifestyle: does anatomy change during the seasonal habitat switches? J Anat 228:757–770.
180. Held LI. 2009. Quirks of human anatomy - an Evo-Devo look at the human body. Cambridge University Press, Cambridge.
181. Henning BG. 2013. Of termites and men: on the ontology of collective individuals. In Beyond Mechanism: Putting Life Back into Biology (Henning BG, Scarfe AC, eds). Lexington Books, Lexington. p. 233–250.
182. Henning BG, Scarfe AC, eds. 2013. Beyond Mechanism: Putting Life Back into Biology. Lexington Books, Lexington.
183. Herrel A, Vanhooydonck B, Porck J, Irschick D. 2008. Anatomical basis of differences in locomotor behavior in Anolis lizards: a comparison between two ecomorphs. Bull Mus Comp Zool 159:213–238.
184. Hinde RA. 1966. Animal behavior: a synthesis of ethology and comparative psychology. McGraw-Hill, New York.
185. Hodin J. 2000. Plasticity and constraints in development and evolution. J Exp Zool B (Mol Dev Evol) 288:1–20.
186. Hoekstra HE, Hoekstra JM, Berrigan D, Vignieri SN, Hoang A, Hill CE, Beerli P, Kingsolver JG. 2001. Strength and tempo of directional selection in the wild. PNAS 98:9157–9160.
187. Hoffmeyer J. 2013. Why do we need a semiotic understanting of life. In Beyond Mechanism: Putting Life Back into Biology (Henning BG, Scarfe AC, eds). Lexington Books, Lexington. p. 147–168.
188. Hoffmeyer J, Kull K. 2003. Baldwin and biosemiotics: what intelligence is for. In Evolution and Learning: The Baldwin Effect Reconsidered (Weber BH, Depew DJ). MIT Press, Cambridge. p. 253–272.
189. Holekamp KE, Swanson EM, Van Meter PE. 2013. Developmental constraints on behavioural flexibility. Philos Trans R Soc Lond B 368:20120350.
190. Holland LZ. 2014. Genomics, evolution and development of amphioxus and tunicates: the Goldilocks principle. J Exp Zool B (Mol Dev Evol) 324:342–352.
191. Holmes SJ. 1944. Recapitulation and its supposed causes. Q Rev Biol 19:319–331.
192. Hone DWE, Benton MJB. 2005. The evolution of large size: how does Cope's Rule work? TRENDS in Ecol & Evol 20:4–6.
193. Hopkins MJ. 2016. Magnitude vs. direction of change and the contribution of macroevolutionary trends to morphological disparity. Biol J Linn Soc 118:116–130.
194. Hudson PE, Corr SA, Payne-Davis RC, Clancy SN, Lane E, Wilson AM. 2011. Functional anatomy of the cheetah (Acinonyx jubatus) hindlimb. J Anat 218:363–374.
195. Hulsey CD, Hollingsworth PR. 2011. Do constructional constraints influence cyprinid (Cyprinidae: Leuciscinae) craniofacial evolution? Biol J Linn Soc 103:136–146.
196. Humphrey N, Skoyles JR, Keynes R. 2005. Human hand-walkers: five siblings who never stood up. DP 77/05. Centre for Philosophy of Natural and Social Science, London School of Economics and Political Science, London, UK.
197. Irie N, Kuratani S. 2011. Comparative transcriptome analysis reveals vertebrate phylotypic period during organogenesis. Nat Commun 2:248.
198. Jablonka E, Lamb MJ. 2005. Evolution in four dimensions - genetic, epigenetic, behavioral, and symbolic variation in the history of life. MIT Press, Cambridge.
199. Jackson FLC, Niculescu MD, Jackson RT. 2013. Conceptual shifts needed to understand the dynamic interactions of genes, environment, epigenetics, social processes, and behavioral choices. Am J Public Health 103:S33–S42.
200. Jacob F. 1977. Evolution and tinkering. Science 196:1161–1166.

201. Jennions M, Kokko H. 2010. Sexual selection. In Evolutionary behavioral ecology (Westneat DF, Fox CW, eds.). Oxford University Press, New York. p. 343–364.

202. Johannsen W. 1911 The `genotype-conception', as I have called the modern view of heredity. Am Naturalist 45:129–159.

203. Johnson NA, Lahti DC, Blumstein DT. 2012. Combating the assumption of evolutionary progress: Lessons from the decay and loss of traits. Evol: Educ & Outreach 5:128–138.

204. Jones KE, Goswami A. 2010. Quantitative analysis of the influences of phylogeny and ecology on phocid and otariid pinniped (Mammalia; Carnivora) cranial morphology. J Zool 280:297–308.

205. Kaji T, Keiler J, Bourguignon T, Miura T. 2016. Functional transformation series and the evolutionary origin of novel forms: evidence from a remarkable termite defensive organ. Evol & Dev 18:78–88.

206. Kathirithamby J, Delgado JA, Collantes F. 2015. Order Strepsiptera. Revista IDE@-SEA 62B:1–10.

207. Kauffman SA. 2013. Evolution beyond Newton, Darwin, and entailing law. In Beyond mechanism: putting life back into biology (Henning BG, Scarfe AC, eds). Lexington Books, Lexington. p. 1–24.

208. Kelley ST, Farrell BD. 1998. Is specialization a dead end? The phylogeny of host use in Dendroctonus bark beetles (Scolytidae). Evolution 52: 1731–1743.

209. Kemp TS. 2016. The origin of higher taxa - paleobiological, developmental and ecological perspectives. Oxford University Press, Oxford.

210. Koestler A. 1967. The ghost in the machine. Macmillan, New York.

211. Kollar EJ, Fisher C. 1980. Tooth induction in chick epithelium: expression of quiescent genes for enamel synthesis. Science 207:993–995.

212. Koyabu D, Son NT. 2014. Patterns of postcranial ossification and sequence heterochrony in bats: life histories and developmental trade-offs. J Exp Zool (Mol Dev Evol) 322B:607–618.

213. Koyabu D, Werneburg I, Morimoto N, Zollikofer CPE, Forasiepi AM, Endo H, Kimura J, Ohdachi SD, Truong SN, Sánchez-Villagra MR. 2014. Mammalian skull heterochrony reveals modular evolution and a link between cranial development and brain size. Nat Commun 5: 3625.

214. Kukalová-Peck J. 1985. Ephemeroid wing venation based upon new gigantic Carboniferous mayflies and basic morphology, phylogeny, and metamorphosis of pterygote insects (Insecta, Ephemerida). Can J Zool 63:933–955.

215. Kull K. 2014. Adaptive evolution without natural selection. Biol J Linn Soc 112:287–294.

216. Laland KN, Galef BG, eds. 2009. The question of animal culture. Harvard, Cambridge.

217. Laland KN, Odling-Smee J, Turner S. 2014. The role of internal and external constructive processes in evolution. J Physiol 592:2413–2422.

218. Laland KN, Uller T, Feldman MW, Sterelny K, Müller GB, Moczek A, Jablonka E, Odling-Smee J. 2015. The extended evolutionary synthesis: its structure, assumptions and predictions. Proc R Soc Lon B 282:10.1098/rspb.2015.1019.

219. Laland K, Matthews B, Feldman MW. 2016. An introduction to niche construction theory. Evol Ecol 30:191–202.

220. Larsen EW. 2004. A view of phenotypic plasticity from molecules to morphogenesis. In Environment, Development and Evolution: Toward a Synthesis (The Vienna Series in Theoretical Biology) (Hall BK, Pearson RD, Müller GB, eds.). A Bradford Book, Massachusetts. p. 118–124.

221. Law CJ, Dorgan KM, Rouse GW. 2014. Relating divergence in polychaete musculature to different burrowing behaviors: a study using Opheliidae (Annelida). J Morphol 275:548–571.

222. Leroi AM. 2003. Mutants: on the form, varieties and errors of the human body. Harper Collins, London.

223. Leroi AM. 2014. The lagoon: how Aristotle invented science. Bloomsbury, London.

224. Levin DA. 2002. The role of chromosomal change in plant evolution. Oxford University Press, New York.
225. Levinton JS. 2001. Genetics, Paleontology and Macroevolution. Cambridge University Press, New York.
226. Levis NA, Pfennig DW. 2016. Evaluating 'plasticity-first' evolution in nature: key criteria and empirical approaches. Trends Ecol Evol 31:563–574.
227. Lewis WH. 1910. The development of the muscular system. In Manual of embryology, Vol 2. (Keibel F, Mall FP, eds.). JB Lippincott, Philadelphia. p. 455–522.
228. Lewontin R. 1995. Genes, Environment, and Organisms. In Hidden histories of science (Silver S, ed.). New York Review Book, New York. p. 115–140.
229. Linde M, Boughner JC, Santana S, Diogo R. 2016. Are more diverse parts of the mammalian skull more labile? Ecol & Evol 6:2318–2324.
230. Lloyd EA. 2015. Adaptationism and the logic of research questions: how to think clearly about evolutionary causes. Biol Theory 10:343–362.
231. Lloyd Morgan C. 1896. Habit and Instinct. Arnold, London.
232. Lindholm M. 2015. DNA dispose, but subjects decide -learning and the extended synthesis. Biosemiotics 8:4431–4461.
233. Lister AM. 2014. Behavioral leads in evolution: evidence from the fossil record. Biol J Linn Soc 112:315–331.
234. Lofeu L, Kohlsdorf T. 2015. Mais que seleção: o papel do ambiente na origem e evolução da diversidade fenotípica. Genética na Escola 10:10–19.
235. Love AC. 2010. Idealization in evolutionary developmental investigation: a tension between phenotypic plasticity and normal stages. Philos Trans R Soc Lond B 365:679–690.
236. Lovejoy AO. 1936. The great chain of being: A study of the history of an idea. Harvard University Press, Cambridge.
237. Lovtrup S. 1978. On von Baerian and Haeckelian recapitulation. Syst Zool 27:348–352.
238. Lowe JW. 2015. Managing variation in the investigation of organismal development: problems and opportunities. Hist Philos Life Sci 237:449–473.
239. Mabee PM. 1993. Phylogenetic interpretation of ontogenic change: sorting out the actual and artefactual in an empirical case study of centrarchid fishes. Zool J Linn Soc 107:175–291.
240. Magnus LZ, Cáceres N. 2016. Phylogeny explains better than ecology or body size the variation of the first lower molar in didelphid marsupials. Mammalia. In press.
241. Mancuso S, Viola A. 2015. Brilliant green: the surprising history and science of plant intelligence. Island Press, Washington DC.
242. Malashichev YB, Rogers LJ, eds. 2002. Behavioral and Morphological Asymmetries in Amphibians and Reptiles, Special issue. Laterality 7:195–295.
243. Margulis L, Sagan D. 1995. What is life? Simon and Shuster, New York.
244. Marroig G, Cheverud JM. 2005. Size as a line of least evolutionary resistance: diet and adaptive morphological radiation in new world monkeys. Evolution 59:1128–1142.
245. Matsuda R. 1987. Animal evolution in changing environments: with special reference to abnormal metamorphosis. Wiley, New York.
246. Matthews B, Aebischer T, Sullam KE, Lundsgaard-Hansen B, Seehausen O. 2016 Experimental evidence of an eco-evolutionary feedback during adaptive divergence. Curr Biol 26:483–489.
247. Mayr E. 1960. The emergence of evolutionary novelties. In: Evolution after Darwin, Vol I. (Tax S, ed.) University of Chicago Press, Chicago. p. 349–380.
248. Mayr E. 1976. Evolution and the diversity of life: selected essays. Harvard University Press, Cambridge.
249. Mayr E. 1988. The probability of extraterrestrial intelligent life. In Toward a new philosophy of biology: observations of an evolutionist (Mayr E, ed). Harvard University Press, Cambridge. p. 67–74.

250. McBrayer LD. 2004. The relationship between skull morphology, biting performance and foraging mode in Kalahari lacertid lizards. Zool J Linn Soc 140:403–416.

251. McFadden J, Al-Khalili J. 2014. Life on the edge - the coming of age of quantum biology. Crown Publishers, New York.

252. Mcnamara KJ. 1986. The role of heterochrony in the evolution of Cambrian trilobites. Biol Rev 61:121–156.

253. McShea DW. 1991. Complexity and evolution: what everybody knows. Biol & Philos 6:303–324.

254. McShea DW. 1996. Metazoan complexity and evolution: is there a trend? Evolution 50:477–492.

255. McShea DW. 2012. Upper-directed systems: a new approach to teleology in biology. Biol & Philos 27:663–684.

256. McShea DW, Brandon RN. 2010. Biology's first law: the tendency for diversity and complexity to increase in evolutionary systems. The University of Chicago Press, Chicago.

257. Meredith RW, Zhang G, Gilbert MT, Jarvis ED, Springer MS. 2014. Evidence for a single loss of mineralized teeth in the common avian ancestor. Science 346:1254390.

258. Minelli A. 2003. The development of animal form: ontogeny, morphology, and evolution. Cambridge University Press, New York.

259. Minelli A. 2009. Forms of becoming - the evolutionary biology of development. Princeton University Press, Princeton.

260. Miyashita T, Diogo R (2016). Evolution of serial patterns in the vertebrate pharyngeal apparatus and paired appendages via assimilation of dissimilar units. Frontiers Ecol Evol - Evo Devo 4:71.

261. Moen DS, Morlon H, Wiens JJ. 2016. Testing convergence vs. history: convergence dominates phenotypic evolution for over 150 million years in frogs. Syst Biol 65:146–60.

262. Molles MC Jr. 2008. Ecology: concepts and applications, 4th Ed. McGraw-Hill, New York.

263. Moore C. 2003. Evolution, development, and the individual acquisition of traits: what we've learned since Baldwin. In Evolution and Learning: The Baldwin Effect Reconsidered (Weber BH, Depew DJ). MIT Press, Cambridge. p. 115–139.

264. Moran LA. 2006. Evolution by accident. http://bioinfo.med.utoronto.ca/Evolution_by_Accident.

265. Morris MRJ. 2014. Plasticity-mediated persistence in new and changing environments. Int J Evol Biol 2014:1–18.

266. Muri RM. 2016. Cortical control of facial expression. J Comp Neurol 524:1578–1585.

267. Narita Y, Kuratani S. 2005. Evolution of the vertebral formulae in mammals: a perspective on developmental constraints. J Exp Zool (Mol Dev Evol) 304B:91–106.

268. Nee S. 2005. The great chain of being. Nature 435:429–429.

269. Neufeld CJ, Palmer AR. 2011. Learning, developmental plasticity, and the rate of morphological evolution. In Epigenetics: linking genotype and phenotype in development and evolution (Hallgrímsson B, Hall BK, eds.). University of California Press, Berkeley. p. 337–356.

270. Newman SA. 2014. Form and function remixed: developmental physiology in the evolution of vertebrate body plans. J Physiol 592:2403–2412.

271. Nijhout HF. 2003. The development and evolution of adaptive polyphenisms. Evol & Dev 5:9–18.

272. Noble D (2006). The Music of Life: Biology Beyond the Genome. OUP, Oxford.

273. Ninova M, Ronshaugen M, Griffiths-Jones S. 2014. Conserved temporal patterns of microRNA expression in Drosophila support a developmental hourglass model. Genome Biol Evol 6:2459–2467.

274. Nonacs P, Blumstein DT. 2010. Predation risk and behavioral life history. In Evolutionary behavioral ecology (Westneat DF, Fox CW, eds.). Oxford University Press, New York. p. 207–221.

275. Ochoa C, Rasskin-Gutman D. 2015. Evo–devo mechanisms underlying the continuum between homology and homoplasy. J Exp Zool (Mol Dev Evol) 324:91–103.
276. Odling-Smee FJ, Laland KN, Feldman MW. 2003. Niche construction – the neglected process in evolution (Monographs in population biology 37). Princeton University Press, Princeton.
277. Olberding JP, Herrel A, Higham TE, Garland T. 2016. Limb segment contributions to the evolution of hind limb length in phrynosomatid lizards. Biol J Linn Soc 117:775–795.
278. Olson ME. 2012. The developmental renaissance in adaptationism. TREE 27:278–287.
279. Olson ME, Arroyo-Santos A. 2015. How to study adaptations (and why to do it that way). Q Rev Biol 90:167–191.
280. O'Malley MA, Koonin EV. 2011. How stands the tree of life a century and a half after The Origin? Biol Direct 6:32.
281. O'Malley MA, Wideman JG, Ruiz-Trillo I. 2016. Losing complexity: the role of simplification in macroevolution. Trends Ecol Evol 31:608–621.
282. Omland KE, Cook LG, Crisp MD. 2008. Tree thinking for all biology: the problem with reading phylogenies as ladders of progress. BioEssays 30:854–867.
283. Ohnishi K, Takahashi A, Tanaka H, Ohnishi T. 1996. Relationship between frequency of upside-down posture and space size around upside-down catfish, Synodontis nigriventris. Biol Sci Space 10:247–251.
284. Opitz JM, Reynolds JF. 1985. The developmental field concept. Am J Med Genet 21:1–11.
285. Ornish D, Lin J, Chan JM, Epel E, Kemp C, Weidner G, Marlin R, Frenda SJ, Magbanua MJ, Daubenmier J, Estay I, Hills NK, Chainani-Wu N, Carroll PR,Blackburn EH. 2013. Effect of comprehensive lifestyle changes on telomerase activity and telomere length in men with biopsy-proven low-risk prostate cancer: 5-year follow-up of a descriptive pilot study. Lancet Oncol 14:1112–1120.
286. Osborn HF. 1897. Organic Selection. Science 15:583–587.
287. Osborn HF. 1929. The Titanotheres of Ancient Wyoming, Dakota, and Nebraska, United States. Geological Survey Monograph 55.
288. Palmer C. 1989. Rape in nonhuman animal species: definitions, evidence, and implications. J Sex Res 26:355–374.
289. Panchen AL. 2001. Étienne Geoffroy St.-Hilaire: father of 'evo-devo'? Evol & Dev 3:41–46.
290. Perez SI, Klaczko J, Rocatti G, Dos Reis SF. 2011. Patterns of cranial shape diversification during the phylogenetic branching process of New World monkeys (Primates: Platyrrhini). J Evol Biol 24:1826–1835.
291. Perez-Barberia FJ, Gordon IJ. 1999. The functional relationship between feeding type and jaw and cranial morphology in ungulates. Oecologia 118:157–165.
292. Pettigrew JB. 1908. Design in nature illustrated by spiral and other arrangements in the inorganic and organic kingdoms as exemplified in matter, force, life, growth, rhythms, &c., especially in crystals, plants, and animals. Longman, Green, and co., London.
293. Pigliucci M, Müller GB, eds. (2010). Evolution - the extended synthesis. MIT Press, Cambridge.
294. Pilcher HR. 2004. Bamboo under extinction threat. Nature 10.1038: news040510-2.
295. Plucain J, Suau A, Cruveiller S, Médigue C, Schneider D, Le Gac M. 2016. Contrasting effects of historical contingency on phenotypic and genomic trajectories during a two-step evolution experiment with bacteria. BMC Evol Biol 16:1.
296. Poe S. 2005. A study of the utility of convergent characters for phylogeny reconstruction: do ecomorphological characters track evolutionary history in Anolis lizards? Zoology 108:337–343.
297. Prabhakar S, Visel A, Akiyama JA, Shoukry M, Lewis KD, Holt A, Plajzer-Frick I, Morrison H, Fitzpatrick DR, Afzal V, Pennacchio LA, Rubin EM, Noonan JP. 2008. Human-specific gain of function in a developmental enhancer. Science 321:1346–1349.

298. Pratarelli ME, Chiarelli B. 2007. Extinction and overspecialization: the dark side of human innovation. Mankind Quart 48:83–98.
299. Previc F. 1991. A general theory concerning the prenatal origins of cerebral lateralization in humans. Psycholog Rev 98:299–334.
300. Puttonen E, Briese C, Mandlburger G, Wieser M, Pfennigbauer M, Zlinszky A, Pfeifer N. 2016. Quantification of overnight movement of birch (Betula pendula) branches and foliage with short interval terrestrial laser scanning. Frontiers in plant science 7: doi: 10.3389/fpls.2016.00222
301. Quint M, Drost H-G, Gabel A, Ullrich KK, Boenn M, Grosse I. 2012. A transcriptomic hourglass in plant embryogenesis. Nature 490: 98–101.
302. Raff RA. 1996. The shape of life: genes, development, and the evolution of animal form. University of Chicago Press, Chicago.
303. Rajakumar R, San Mauro D, Dijkstra MB, Huang MH, Wheeler DE, Hiou-Tim F, Khila A, Cournoyea M, Abouheif E. 2012. Ancestral developmental potential facilitates parallel evolution in ants. Science 335:79–82.
304. Reid RGB. 2004. Epigenetics and environment: the historical matrix of Matsuda's pan-environmentalism. In Environment, development and evolution: toward a synthesis (The Vienna Series in Theoretical Biology) (Hall BK, Pearson RD, Müller GB, eds). A Bradford Book, Massachusetts. p. 8–35.
305. Reiss JO. 2009. Not by design: retiring Darwin's watchmaker. University of California Press, Berkeley.
306. Rice SH. 1998. The evolution of canalization and the breaking of von Baer's laws: modeling the evolution of development with epistasis. Evolution 52:647–656.
307. Richards RJ. 1992. The meaning of evolution: the morphological construction and ideological reconstruction of Darwin's theory. University of Chicago Press, Chicago.
308. Richards RJ. 2002. The romantic conception of life: science and philosophy in the age of Goethe. University of Chicago Press, Chicago.
309. Richards RJ. 2008. The tragic sense of life: Ernst Haeckel and the struggle over evolutionary thought. University of Chicago Press, Chicago.
310. Rigato E, Minelli A. 2013. The great chain of being is still here. Evo: Educ & Outreach 6:18.
311. Roe A, Simpson GG, eds. 1958. Behavior and evolution. Yale University Press, New Haven.
312. Rolian C, Lieberman DE, Hallgrímsson B. 2010. The coevolution of human hands and feet. Evolution 64:1558–1568.
313. Romero A. 2001. The biology of hypogean fishes. Kluwer Academic Publishers, London.
314. Rosslenbroich B. 2006. The notion of progress in evolutionary biology - the unresolved problem and an empirical suggestion. Biol & Philos 21:41–70.
315. Rostand J. 1964. Etienne Geoffroy Saint-Hilaire et la tératogénèse expérimentale. Rev Hist Sci 17:41–50.
316. Rundle HD, Boughman JW. 2010. Behavioral ecology and speciation. In Evolutionary behavioral ecology (Westneat DF, Fox CW, eds.). Oxford University Press, New York. p. 471–487.
317. Ruth A, Meindl RS, Raghanti MA, Lovejoy CO. 2016. Locomotor pattern fails to predict foramen magnum angle in marsupials, rodents, and strepsirrhine primates. J Hum Evol 94:45–52.
318. Ruse M. 1996. Monad to man: the concept of progress in evolutionary biology. Harvard University Press, Cambridge.
319. Ruse M. 2003. Darwin and design - does evolution have a purpose? Harvard University Press, Cambridge.
320. Ruse M. 2013. From organisms to mechanisms - and halfway back? In Beyond Mechanism: Putting Life Back into Biology (Henning BG, Scarfe AC, eds). Lexington Books, Lexington. p. 409–430.

321. Sapolsky RM. 2005. The influence of social hierarchy on primate health. Science 308:648–652.

322. Sakai T, Hirata S, Fuwa K, Sugama K, Kusunoki K, Makishima H, Eguchi T, Yamada S, Ogihara N, Takeshita H. 2012. Fetal brain development in chimpanzees vs. humans. Curr Biol 22:R791–792.

323. Santana SE, Dobson SD, Diogo R. 2014. Plain faces are more expressive: comparative study of facial color, mobility and musculature in primates. Biol Letters 10:20140275.

324. Saunders WB, Work DM, Nikolaeva SV (1999) Evolution of complexity in Paleozoic ammonoid sutures. Science 286:760–763.

325. Scarfe AC (2013). On the ramifications of the theory of organic selection for environmental and evolutionary. In Beyond Mechanism: Putting Life Back into Biology (Henning BG, Scarfe AC, eds). Lexington Books, Lexington. p. 259–286.

326. Scheider L, Liebal K, Oña L, Burrows A, Waller B. 2014. A comparison of facial expression properties in five hylobatid species. Am J Primatol 76:618–628.

327. Schlichting CD, Pigliucci M. 1998. Phenotypic evolution: a reaction norm perspective. Sinauer Associates, Sunderland.

328. Schmalhausen II. 1949. Factors of evolution. Blakiston. Philadelphia.

329. Schulte JA II, Losos JB, Cruz FB, Núñez H. 2004. The relationship between morphology, escape behavior, and microhabitat occupation in the iguanid lizard genus Liolaemus. J Evol Biol 17:48–420.

330. Schwenk K, Wagner GP. 2003. Constraint. In Keywords in evolutionary developmental biology (Hall BK, Olson W, eds.). Harvard University Press, Cambridge. p 52–61.

331. Sears K, Capellini T, Diogo R. 2015. On the serial homology of the pectoral and pelvic girdles of tetrapods. Evolution 69:2543–2555.

332. Seger J, Stubblefield J. 1996. Optimization and adaptation. In Adaptation (Rose MR, Lauder GV, eds.). San Diego, Academic Press. p. 93–124.

333. Senter P, Moch JG. 2015. A critical survey of vestigial structures in the postcranial skeletons of extant mammals. PeerJ 3:e1439.

334. Shapiro BL, Hermann J, Opitz JM. 1983. Down syndrome - a disruption of homeostasis. Am J Med Genet 14:241–269.

335. Shaw KL, Wiley C. 2010. The genetic basis of behavior. In Evolutionary behavioral ecology (Westneat DF, Fox CW, eds.). Oxford University Press, New York. p. 71–74.

336. Shkil F, Smirnov SV. 2015. Experimental approach to the hypotheses of heterochronic evolution in lower vertebrates. Paleontolog J 49:1624–1634.

337. Simoes PM, Ott SR, Niven JE. 2016. Environmental adaptation, phenotypic plasticity, and associative learning in insects: the desert locust as a case study. Integr Comp Biol 56:914–924.

338. Simpson GG. 1953. The Baldwin effect. Evolution 2:110–117.

339. Sinervo B, Svensson EI. 2004. The origin of novel phenotypes: correlational selection, epistasis, and speciation. In Environment, development and evolution: toward a synthesis (The Vienna Series in Theoretical Biology) (Hall BK, Pearson RD, Müller GB, eds). A Bradford Book, Massachusetts. p. 171–194.

340. Singh JAL, Zingg RM, Feuerbach PJA. 1966. Wolf-children and feral man. Archon Books, Hamden.

341. Slaby O. 1990. On the validity of von Baer's laws in evolutionary morphology. Folia morphologica 38: 293–300.

342. Smith CM, Molnar JL, Ziermann JM, Gondre-Lewis M, Sandone C, Aziz AM, Bersu ET, Diogo R. 2015. Muscular and skeletal anomalies in human trisomy in an Evo-Devo context: description of a T18 cyclopic fetus and comparison between Edwards (T18), Patau (T13) and Down (T21) syndromes using 3-D imaging and anatomical illustrations. Taylor & Francis, Oxford.

343. Smith KK. 2006. Craniofacial development in marsupial mammals: developmental origins of evolutionary change. Dev Dyn 235:1181–1193.

344. Smith RJ. 2016. Freud and evolutionary anthropology's first just-so story. Evol Anthropol 25:50–53.
345. Snyder-Mackler N, Remon JS, Kohn J, Brinkworth J, Johnson Z, Wilson M, Barreiro L, Tung J. 2016. The genomic signature of social adversity in rhesus macaques. Oral presentation at ISEMPH2016: http://easychair.org/smart-program/ISEMPH2016/.
346. Standen EM, Du RY, Larsson HCE. 2014. Developmental plasticity and the origin of tetrapods. Nature 513:54–58.
347. Stanley SM. 1979. Macroevolution - patterns and process. W.H. Freeman & Co.,
348. Stearns SC, Medzhitov R. 2015. Evolutionary medicine. Sinauer Associates, Sunderland.
349. Streelman JT, Danley PD. 2003. The stages of vertebrate evolutionary radiation. TREE 18:126–131.
350. Sultan SE. 2016. Organisms & environment - ecological development, niche construction, and adaptation. Oxford University Press, Oxford.
351. Suraci JP, Clinchy M, Dill LM, Roberts D, Zanette LY. 2016. Fear of large carnivores causes a trophic cascade. Nat Commun 7:10698.
352. Thompson DB. 1992. Consumption rates and the evolution of diet-induced plasticity in the head morphology of Melanoplus femurrubrum (Orthoptera: Acrididae). Oecologia 89:204–213.
353. Thorpe WH. 1956. Learning and instinct in animals. Methuen, London.
354. Tinbergen N. 1953. The herring gull's world. Collins, London.
355. Tubbs RS, Malefant J, Loukas M, Jerry Oakes W, Oskouian RJ, Fries FN. 2016. Enigmatic human tails: a review of their history, embryology, classification, and clinical manifestations. Clin Anat 29:430–438.
356. Tulli MJ, Cruz FB, Herrel A, Vanhooydonck B, Abdala V. 2009. The interplay between claw morphology and habitat use in neotropical iguanian lizards. Zoology 112:379–392.
357. Tulli MJ, Abdala V, Cruz FB. 2011. Relationships among morphology, clinging performance and habitat use in Liolaemini lizards. J Evol Biol 24:843–855.
358. Tulli MJ, Herrel A, Vanhooydonck B, Abdala V. 2012. Is phylogeny driving tendon length in lizards? Acta Zool 93:319–329.
359. Turner JS. 2000. The extended organism - the physiology of animal-built structures. Harvard University Press, Cambridge.
360. Turner JS. 2007. The tinkerer's accomplice: how design emerges from life itself. Harvard University Press, Cambridge.
361. Turner JS. 2013. Biology's second law: homeostasis, purpose and desire. In Beyond Mechanism: Putting Life Back into Biology (Henning BG, Scarfe AC, eds). Lexington Books, Lexington. p. 183–204.
362. Turner JS. 2016. Homeostasis and the physiological dimension of niche construction theory in ecology and evolution. Evol Ecol 30:203–219.
363. Valles SA. 2011. Evolutionary medicine at twenty: rethinking adaptationism and disease. Biol & Philos 27:241–261.
364. Vanhooydonck B, Van Damme R. 1999. Evolutionary relationships between body shape and habitat use in lacertid lizards. Evol Ecol Res 1:785–805.
365. Velo-Antón G, Zamudio KR, Cordero-Rivera A. 2012. Genetic drift and rapid evolution of viviparity in insular fire salamanders (Salamandra salamandra). Heredity 108:410–418.
366. Verhulst J. 2003. Developmental dynamics in humans and other primates: discovering evolutionary principles through comparative morphology. Adonis, Ghent.
367. Vermeij GJ. 1973. Biological versatility and earth history. PNAS 70:1936–1938.
368. Vidal-García M, Byrne PG, Roberts JD, Keogh JS. 2014. The role of phylogeny and ecology in shaping morphology in 21 genera and 127 species of Australo-Papuan myobatrachid frogs. J Evol Biol 27:181–192.
369. Vinicius M. 2012. Modular evolution: how natural selection produces biological complexity. Cambridge University Press, Cambridge.

370. Vitt LJ, Pianka ER. 2005. Deep history impacts present-day ecology and biodiversity. PNAS 102:7877–7881.
371. Vrba ES. 2004. Ecology, development, and evolution: perspectives from the fossil record. In Environment, development and evolution: toward a synthesis (The Vienna Series in Theoretical Biology) (Hall BK, Pearson RD, Müller GB, eds). A Bradford Book, Massachusetts. p. 85–105.
372. Waddington CH. 1942. Canalization of development and the inheritance of acquired characters. Nature 150:563–565.
373. Waddington CH. 1952a. Selection of the genetic basis for an acquired character. Nature 169:278.
374. Waddington CH. 1952b. Selection of the genetic basis for an acquired character: reply to Begg. Nature 169:625–626.
375. Waddington CH. 1953a. The "Baldwin effect", "genetic assimilation" and "homeostasis". Evolution 7:386–387.
376. Waddington CH. 1953b. Epigenetics and evolution. Symp Soc Exp Biol 7:186–199.
377. Waddington CH. 1953c. The evolution of adaptations. Endeavour 12:134–139.
378. Waddington CH. 1953d. Genetic assimilation of an acquired character. Evolution 7:118–126.
379. Waddington CH. 1956. Genetic assimilation of the bithorix phenotype. Evolution 10:1–13.
380. Waddington CH. 1957. The strategy of the genes. George Allen & Unwin Ltd., London.
381. Waddington CH. 1959. Canalization of development and genetic assimilation of acquired characters. Nature 183:1654–1655.
382. Waddington CH. 1961a. Genetic assimilation. Adv Genet 10:257–290.
383. Waddington CH. 1961b. The nature of life. Harper & Row, New York.
384. Waddington CH. 1975. The evolution of an evolutionist. Cornell University Press, Ithaca.
385. Wagner A. 2014. Arrival of the fittest: solving evolution's greatest puzzle. Oneworld publications, London.
386. Wagner GP. 2014. Homology, genes, and evolutionary innovation. Princeton University Press, Princeton.
387. Wagner GP, Schwenk K. 2000. Evolutionarily stable configurations: functional integration and the evolution of phenotypic stability. In Evolutionary Biology, vol. 31 (Hecht MK, MacIntyre RJ, Clegg MT, eds.). Kluwer Academic/Plenum Press, New York. p. 155–217.
388. Wainwright PC, Alfaro ME, Bolnick DI, Hulsey D. 2005. Many-to-one mapping of form to function: a general principle in organismal design. Integr Comp Biol 45:256–262.
389. Wang X, Clarke JA. 2014. Phylogeny and forelimb disparity in waterbirds. Evolution 68:2847–2860.
390. Weber BH, Depew DJ, eds. (2003). Evolution and learning: the Baldwin effect reconsidered. MIT Press, Cambridge.
391. Weisbecker V, Nilsson M. 2008. Integration, heterochrony, and adaptation in pedal digits of syndactylous marsupials. BMC Evol Biol 8:160.
392. Weisbecker V, Goswami A, Wroe S, Sánchez-Villagra MR. 2008. Ossification heterochrony in the therian postcranial skeleton and the marsupial-placental dichotomy. Evolution 59:2691–2704.
393. West-Eberhard MJ. 2003. Developmental plasticity and evolution. Oxford University Press, Oxford.
394. West-Eberhard MJ. 2004. Ryuichi Matsuda: a tribute and a perspective on pan-environmentalism and genetic assimilation. In Environment, development and evolution: toward a synthesis (The Vienna Series in Theoretical Biology) (Hall BK, Pearson RD, Müller GB, eds). A Bradford Book, Massachusetts. p. 109–116.
395. West-Eberhard MJ. 2007. Dancing with DNA and flirting with the ghost of Lamarck. Biol & Philos 22:439–451.
396. West-Eberhard MJ. 2014. Darwin's forgotten idea: the social essence of sexual selection. Neurosci Biobehav Rev 46:501–508.

397. Westneat DF, Fox CW, eds. 2010. Evolutionary behavioral ecology. Oxford University Press, New York.
398. Whyte LL. 1965. Internal factors in evolution. Braziller, New York.
399. Wiedersheim R. 1895. The structure of man - and index to his past history. MacMillan and Co., London.
400. Wiens JJ. 2011. Re-evolution of lost mandibular teeth in frogs after more than 200 million years, and re-evaluating Dollo's Law. Evolution 65:1283–1296.
401. Wiens JJ, Bonett RM, Chippindale PT. 2005. Ontogeny discombobulates phylogeny: Paedomorphosis and higher-level salamander relationships. Syst Biol 54:91–110.
402. Wiens JJ, Sparreboom M, Arntzen JW. 2011. Crest evolution in newts: implications for reconstruction methods, sexual selection, phenotypic plasticity and the origin of novelties. J Evol Biol 24:2073–2086.
403. Wood J. 1867a. On human muscular variations and their relation to comparative anatomy. J Anat Physiol 1867:44–59.
404. Wood J. 1867b. Variations in human myology observed during the Winter Session of 1866-7 at King's College London. Proc Royal Soc 15:518–545.
405. Wood J. 1868. Variations in human myology observed during the winter session of 1867-68 at King's College, London. Proc Royal Soc Lond 16:483–525.
406. Zaaf A, Herrel A, Aerts P, De Vree F. 1999. Morphology and morphometrics of the appendicular musculature in geckoes with different locomotor habits (Lepidosauria). Zoomorph 119: 9–22.
407. Zanette LY, White AF, Allen MC, Clinchy M. 2011. Perceived predation risk reduces the number of offspring songbirds produce per year. Science 334:1398–1401.
408. Zanno LE, Makovicky PJ. 2011. Herbivorous ecomorphology and specialization patterns in theropod dinosaur evolution. PNAS 108:232–237.
409. Ziermann JM, Diogo R. 2013. Cranial muscle development in the model organism Ambystoma mexicanum: implications for tetrapod and vertebrate comparative and evolutionary morphology and notes on ontogeny and phylogeny. Anat Rec 296:1031–1048.
410. Ziermann JM, Diogo R. 2014. Cranial muscle development in frogs with different developmental modes: direct development vs. biphasic development. J Morphol 275:398–413.
411. Yablokov AV. 1974. Variability of Mammals. Amerind Publishing, New Delhi.
412. Ydenberg RC. 2010. Decision theory. In Evolutionary behavioral ecology (Westneat DF, Fox CW, eds.). Oxford University Press, New York. p. 131–147.
413. Young JL. 2013. The Baldwin effect and the persistent problem of preformation vs. epigenesis. New Ideas Psychol 31:356–362.
414. Young NM, Hu D, Lainoff AJ, Smith F, Diaz R, Tucker AS, Trainor PA, Schneider RS, Hallgrimsson B, Marcucio R. 2014. Embryonic bauplans and the developmental origins of facial diversity and constraint. Development 141:1059–1063.
415. Youson JH. 2004. The impact of environmental and hormonal cues on the evolution of fish metamorphosis. In Environment, development and evolution: toward a synthesis (The Vienna Series in Theoretical Biology) (Hall BK, Pearson RD, Müller GB, eds). A Bradford Book, Massachusetts. p. 240–277.

Index

© Springer International Publishing AG 2017
R. Diogo, *Evolution Driven by Organismal Behavior*,
DOI 10.1007/978-3-319-47581-3

Printed in the United States
By Bookmasters